Modern Electrical Installation
for Craft Students

Modern Electrical Installation
for Craft Students

Volume 3

Second Edition

Brian Scaddan
T.Eng.(C.E.I.), M.I.Elec.I.E.

Heinemann : London

William Heinemann Ltd
10 Upper Grosvenor Street, London W1X 9PA

LONDON MELBOURNE JOHANNESBURG AUCKLAND

First published by Butterworth & Co. (Publishers) Ltd 1983
Reprinted 1986
First published by William Heinemann Ltd 1987

© William Heinemann Ltd 1987

British Library Cataloguing in Publication Data
Scaddan, Brian
   Modern electrical installation for craft srudents. —2nd ed.
   Vol. 3
   1. Electric wiring, Interior
   I. Title
   621.319'24   TK3271

ISBN 0 434 91825 3

Printed and bound by Thomson Litho Ltd, East Kilbride, Scotland

Contents

Preface

1 Communication and Industrial Studies 1

2 Alternating-current Theory 6

3 Machines 36

4 Rectification 83

5 Earthing 91

6 Installation Systems 103

7 Lighting and Illumination 120

8 Testing, Inspection and Instruments 132

Answers to Test Questions 149

Index 151

To my wife

Preface

This series of three books is designed to meet the syllabus requirements of Parts I and II of the City and Guilds 236 course in Electrical Installation Work, Volumes I and II covering Part I of the course, Volume III covering Part II.

Each book deals with craft theory, associated subjects, and a study of the electrical industries. The craft theory and associated subjects are treated, not as separate entities, but as a whole, encouraging the student to understand not just *how* but *why* electrical installations are designed and carried out in particular ways.

Wherever possible, calculations and examples have been devised to have values and wording appropriate to the work of the craft student, and the content of the volumes is comprehensive enough to relieve the student and staff of laborious note-taking. The object here is to encourage discussion and leave more time for demonstrations, individual investigation of the subject matter and classwork.

A great many useful diagrams, together with a simple method of approach, ensure that students will have no difficulty in understanding either the English or the concepts.

The author wishes to express his thanks for all the help, encouragement and advice offered by his colleagues Mr. R.V. Gorman, Mr. C.A. Peach and Mr. R. Bridger.

CHAPTER 1
Communication and Industrial Studies

Just about all working processes are governed by one or more sets of regulations. The use of electricity is no exception to this, the main body of regulations being as follows:

British Standards Specifications (B.S.)

British Standards Codes of Practice (C.P.)

I.E.E. (Institution of Electrical Engineers) Wiring Regulations

Electricity (Factories Acts) Special Regulations 1908 and 1944

Electricity Supply Regulations 1957

Electricity (Factories Acts) Special Regulations

These are statutory regulations (i.e. they may be enforced by law) and are the basis of the I.E.E. Regulations. They lay down, in broad terms, the measures to be taken to ensure the safe installation and use of electrical equipment.

Electricity Supply Regulations

Once again these are statutory regulations. In this case they ensure the safety and welfare of the public and that a proper and sufficient supply of electricity is provided.

I.E.E. Wiring Regulations

The I.E.E. Regulations are not statutory, but are based on the statutory Acts and internationally agreed codes of safety. They lay down recommendations for the safe installation and use of electrical equipment in buildings, and are therefore of great importance to members of the electrical contracting industry.

British Standards Specifications

British Standards (abbreviated to B.S.) ensure a national uniformity in the quality, performance, dimensions and listing of materials.

British Standards Codes of Practice

Codes of Practice (C.P.s) are issued by the British Standards Institution and recommend standards of good practice. In the case of electrical installations, they follow, in general, the recommendations of the I.E.E. Regulations. In some cases, when a choice of methods is available, they select one as preferred practice.

All sources of information mentioned are important to the installation electrician and he or she should be familiar with them all.

Also important is a knowledge of the many different Boards, Associations, Unions and schemes which influence the electrician's work.

Joint Industrial Board

The Joint Industrial Board (J.I.B.) is a national organization which basically acts as a means of liaison between unions and employers in the subject of gradings and rates of pay for employees. Hence a J.I.B.-graded electrician will have attained a certain academic and practical standard and will receive a set basic wage wherever he is working.

Trade Unions

There are several unions to which the electrician can belong. Whichever he chooses, if indeed he has a choice, he should remember that the unions' basic role is to ensure that employees enjoy satisfactory working conditions and rates of pay. They are the employees' voice with which to speak to management, and in the event of any serious dispute with the employer, it is best to let the union deal with the matter in the correct way, using the approved disputes procedure and conciliation machinery.

Electrical Contractors' Association

The Electrical Contractors' Association (E.C.A.) was established in 1901 and requires its members to have a high standard of workmanship.

Anyone employing the services of an E.C.A.-registered firm can expect top-quality work carried out in an efficient manner. Contractors can become members only after their work has been carefully examined by E.C.A. experts.

Contracts and tenders

The majority of electricians are employed by a contracting firm and therefore are not likely to become involved with the administrative side of the business. However, it is important that the electrician is aware of the procedures involved in obtaining the work he is to carry out.

Usually, the first step in obtaining a contract to carry out an installation, is to tender a price for the work.

Tenders

A tender is, by dictionary definition, an offer to supply goods and/or services at a fixed rate. In many cases this is a simple procedure but on larger jobs the tender can become complicated, and considerable experience is necessary to complete such a tender correctly.

Contracts

The law relating to contracts is extremely complicated and involved, and hence only the most basic concepts will be considered.

In simple terms, for any job there is a *main contractor*, which can be an electrical installation firm, building firm or a decorating firm, etc., depending on the work to be done. This main contractor is responsible to the client (that is, the person ordering the work to be done), either directly or via an agent such as an architect.

Should the main contractor employ the services of another firm, this firm is called the *sub-contractor* and is responsible to the main contractor.

A typical sequence of events is as follows:

1. The client approaches an architect with a view to having, say, a hotel designed and built.

2. The architect designs the building and his design is approved by the client.

3. A specification and a bill of quantities are prepared.

4. The work is put out to tender, and eventually one is selected — not necessarily the lowest-priced.

5. The architect may nominate the sub-contractors, i.e. painters, electricians, plumbers, etc., or he may leave it to the main contractor. In any event the sub-contracts will go to tender. This is where the ability to read and interpret drawings, bills of quantities and specifications is so important.

Specifications

Specifications indicate the work to be done and/or the type of material to be used; that is, they specify details regarding the work.

Bills of quantities

Usually prepared by a quantity surveyor, bills of quantities indicate, for each trade concerned with the work, the quantity and sometimes cost of the materials to be used.

Variation order

In the case of electrical installation work, there is every chance, on a big job, that some variation from the programme will occur. It is very important that the site electrician notifies his superiors *immediately* of any change. A *variation order* can then be made out which will enable the new work to be carried out without breaking any of the terms of the contract.

Daywork
It may be necessary to carry out work in addition to that referred to in the contract and this work will be the subject of daywork.

Daywork is normally charged at a higher rate than the work tendered for on the main contract, and the charges are usually quoted on the initial tender.

Typical additional charges are as follows:

120% − labour
20% − materials } added to normal rates
5% − plant

It is important that the members of the installation team on site record on 'daywork forms' all extra time, plant and materials used.

It must always be remembered that employer and employees are both essential for a business to exist. For it to succeed, the two sides must work in harmony. From the point of view of the electrician on site, this involves − apart from doing a good wiring job − accurate recording of time on time sheets and daywork sheets, recording and checking deliveries of materials on site, ensuring that all materials stored on site are safe, and keeping in constant contact with his employers, the main contractor and other sub-contractors.

Bar charts

In order that a job may be carried out in the most efficient manner, some job programmers use a bar chart. This is simply a method of showing graphically each stage of work to be completed on a job. This is best illustrated by an example, though it should be borne in mind that there are several ways of drawing a bar chart, this being only one of the possible methods.

Example 1.1
An electrical contracting firm has the job of rewiring an old three-storey dwelling, each floor of which is to be converted into a self-contained flat. Two pairs of men will do the work and the estimated time for each stage per pair is as follows:

Removal of old wiring per floor − 3 days
Installing new wiring per floor − 1 week
Testing and inspection per floor − 1 day

Communication and industrial studies 5

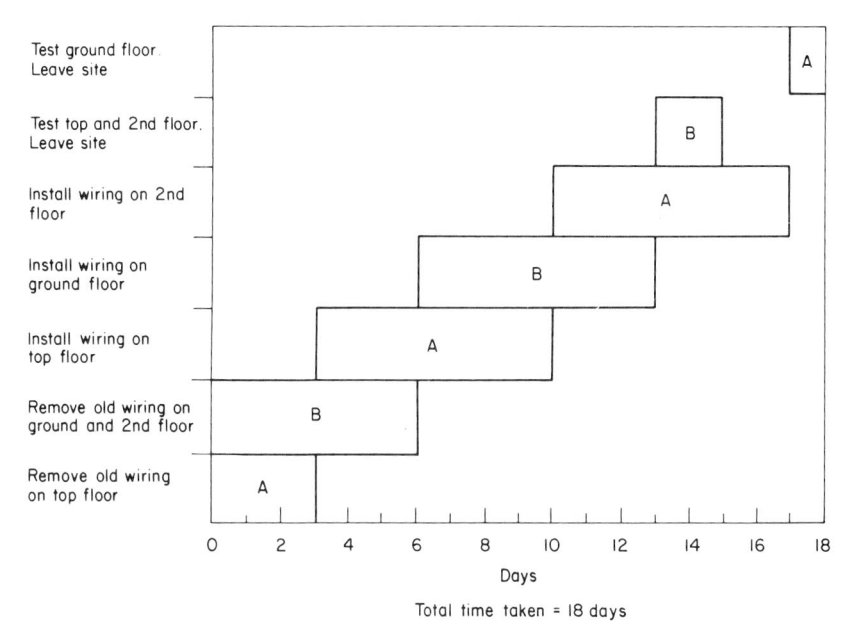

Diagram 1   Bar chart. Group A – two men; group B – two men

By using a bar chart (*Diagram 1*), estimate the least time in which the work may be completed.

Test project

Examine carefully the plans and elevations of the church (pp 146–9) and prepare the following:

(a) a specification for the wiring of the building (choose your own type and manufacture of fittings);

(b) indicate on the drawings the positions of all accessories, using B.S. 3939 symbols (*see* Volume 2);

(c) a requisition of all the materials needed;

(d) a simple bar chart showing the least time for completion assuming two pairs of men.

CHAPTER 2
Alternating-current Theory

As much of this topic has been covered in Volume 2, this chapter will recap on that work and investigate other methods used for solving problems in a.c. circuits.

Pure resistance (*Diagram 2*)

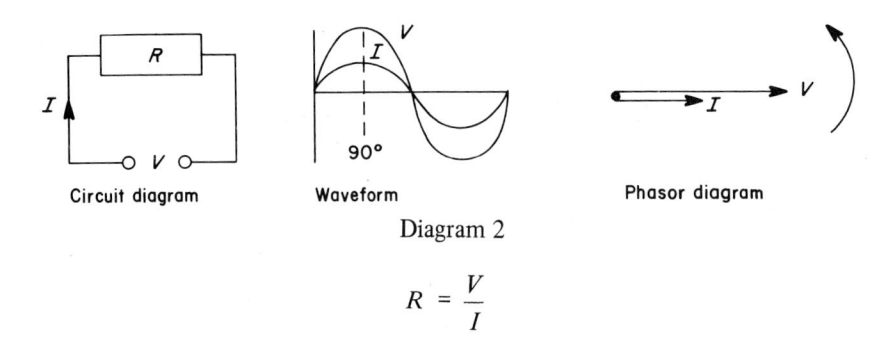

Circuit diagram   Waveform   Phasor diagram

Diagram 2

$$R = \frac{V}{I}$$

Pure inductance (*Diagram 3*)

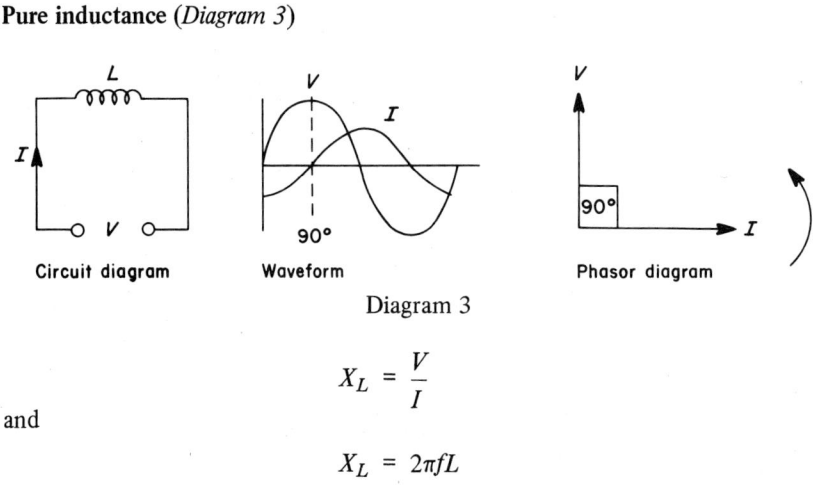

Circuit diagram   Waveform   Phasor diagram

Diagram 3

$$X_L = \frac{V}{I}$$

and

$$X_L = 2\pi f L$$

Alternating-current theory 7

Pure capacitance (*Diagram 4*)

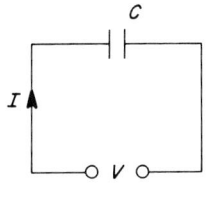

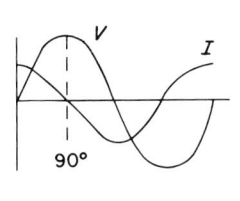

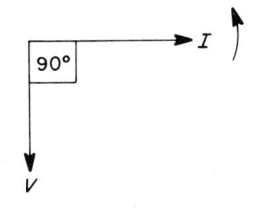

Circuit diagram   Waveform   Phasor diagram

Diagram 4

$$X_C = \frac{V}{I}$$

and

$$X_C = \frac{1}{2\pi f C}$$

R and *L* in series (*Diagram 5*)

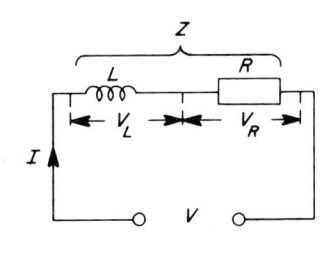

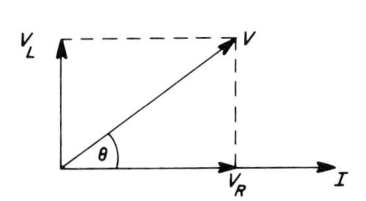

Circuit diagram   Phasor diagram

Diagram 5

$$Z = \frac{V}{I}$$

From the phasor diagram of voltages (*Diagram 5*), an impedance triangle may be formed (*Diagram 6*). By Pythagoras' theorem,

8   Alternating-current theory

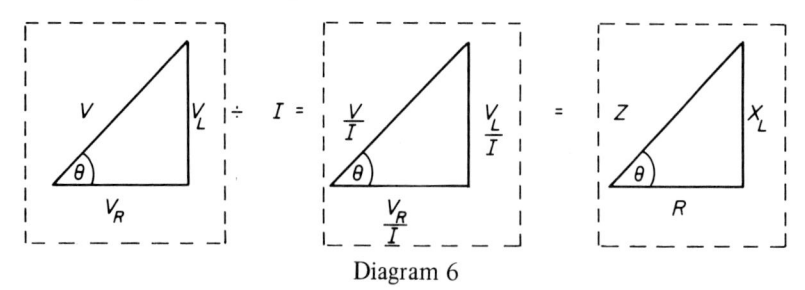

Diagram 6

$$Z = \sqrt{R^2 + X_L^2}$$

Also

$$\cos \theta = \frac{R}{Z} = \text{power factor (P.F.)}$$

R and C in series (*Diagram 7*)

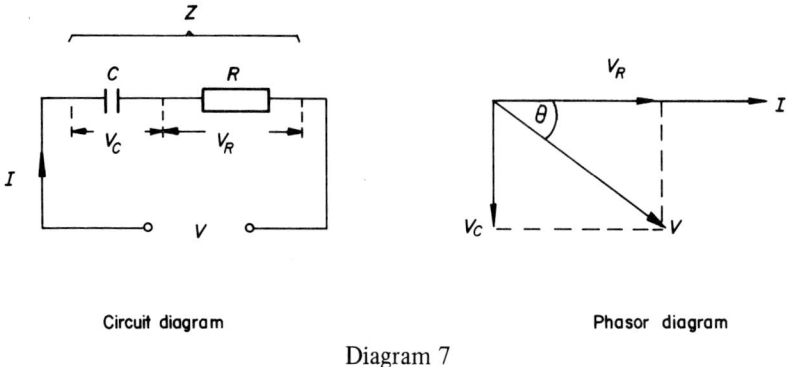

Circuit diagram

Phasor diagram

Diagram 7

It is clear that a similar impedance triangle may be formed, as shown in *Diagram 8*.

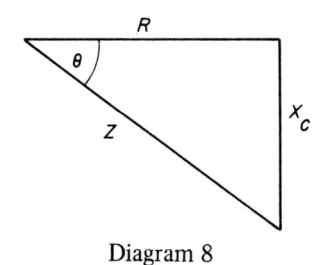

Diagram 8

and
$$Z = \sqrt{R^2 + X_C^2}$$

$$\cos\theta = \frac{R}{Z} = \text{P.F.}$$

R, L and C in series (*Diagram 9*)

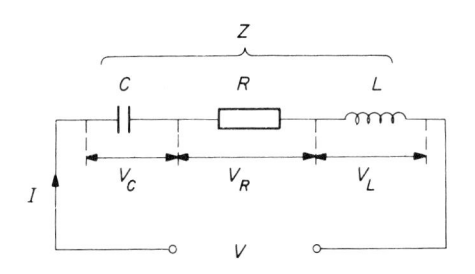

Circuit diagram

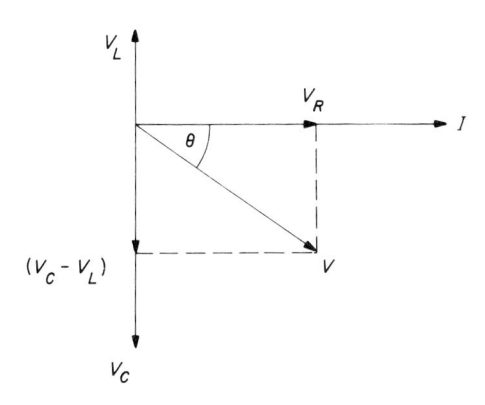

Phasor diagram

Diagram 9 (It is assumed that V_C is greater than V_L)

The impedance triangle will be as shown in *Diagram 10*.

10 Alternating-current theory

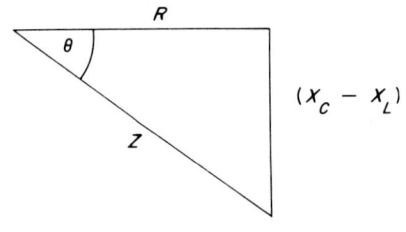

Diagram 10

$$\therefore Z = \sqrt{R^2 + (X_C - X_L)^2}$$

or if V_L is greater than V_C,

$$Z = \sqrt{R^2 + (X_L - X_C)^2}$$

and

$$\cos\theta = \frac{R}{Z} = \text{P.F.}$$

The following problems will be solved using different methods.

Example 2.1

A choke has an inductive reactance of 20 Ω and a resistance of 15 Ω. When connected to an a.c. supply it draws a current of 9.6 A. Determine the value of the supply voltage and the power factor (*Diagram 11*).

Diagram 11

Method 1, using phasors

$$V_L = I \times X_L$$
$$= 9.6 \times 20$$
$$= 192\,\text{V}$$

$$V_R = I \times R$$
$$= 9.6 \times 15$$
$$= 144\,\text{V}$$

Draw V_L and V_R to scale on a phasor diagram (*Diagram 12*).

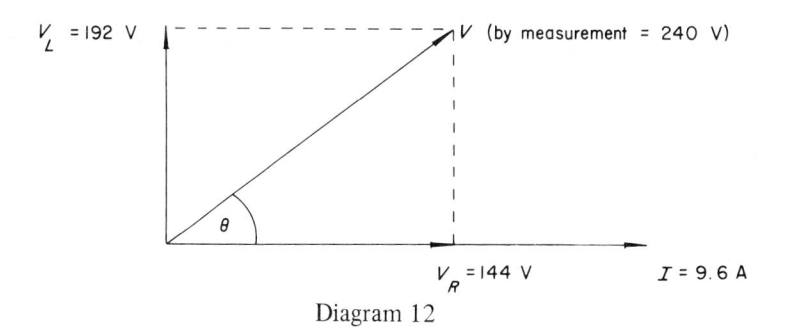

```
V_L = 192 V                    V (by measurement = 240 V)


           θ
                  V_R = 144 V        I = 9.6 A
```
Diagram 12

From *Diagram 12*, $V = 240\,\text{V}$.

Power factor $= \cos \theta$
By measurement $\theta = 53.13°$
From cosine tables, $\cos 53.13° = 0.6$
$\therefore$ P.F. $= 0.6$ lagging

Method 2, using the theorem of Pythagoras (impedance)

$$Z = \frac{V}{I}$$

and

$$\begin{aligned}
Z &= \sqrt{R^2 + X_L^2} \\
&= \sqrt{15^2 + 20^2} \\
&= \sqrt{625} \\
&= 25\,\Omega
\end{aligned}$$

$$\therefore 25 = \frac{V}{9.6}$$

$$\begin{aligned}
\therefore V &= 25 \times 9.6 \\
&= 240\,\text{V}
\end{aligned}$$

12   Alternating-current theory

and

$$\text{P.F.} = \frac{R}{Z}$$
$$= \frac{15}{25}$$
$$= 0.6 \text{ lagging}$$

Method 3, using the theorem of Pythagoras (voltage)
From method 1,

$$V_L = 192\,\text{V} \quad \text{and} \quad V_R = 144\,\text{V}$$

Diagram 13

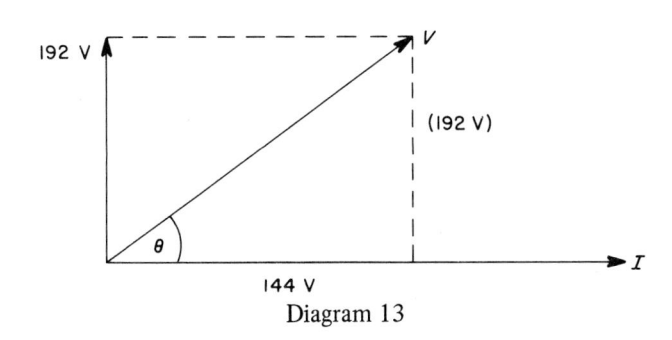

From *Diagram 13*,

$$V^2 = V_L^2 + V_R^2$$
$$\therefore V = \sqrt{V_L^2 + V_R^2}$$
$$= \sqrt{192^2 + 144^2}$$
$$= \sqrt{57\,600}$$
$$= 240\,\text{V}$$

$$\text{P.F.} = \cos\theta = \frac{\text{base}}{\text{hypotenuse}}$$
$$= \frac{144}{240}$$
$$= 0.6 \text{ lagging}$$

Method 4, by trigonometry
From method 1,
$$V_L = 192 \text{ V} \quad \text{and} \quad V_R = 144 \text{ V}$$

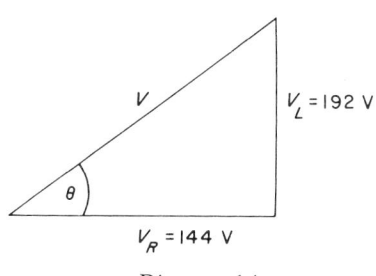

Diagram 14

From *Diagram 14*,
$$\tan \theta = \frac{\text{perpendicular}}{\text{base}} = \frac{V_L}{V_R}$$
$$= \frac{192}{144}$$
$$= 1.333$$

From tangent tables,
$$\theta = 53.13°$$

$$\cos \theta = \frac{\text{base}}{\text{hypotenuse}} = \frac{V_R}{V}$$

$$\therefore V = \frac{V_R}{\cos \theta}$$

From cosine tables,
$$\cos 53.13° = 0.6 \text{ (power factor)}$$

$$\therefore V = \frac{144}{0.6}$$
$$= 240 \text{ V}$$
$$\text{P.F.} = 0.6 \text{ lagging}$$

14   Alternating-current theory

Example 2.2

From the circuit shown in *Diagram 15*, determine the value of the supply voltage and the power factor.

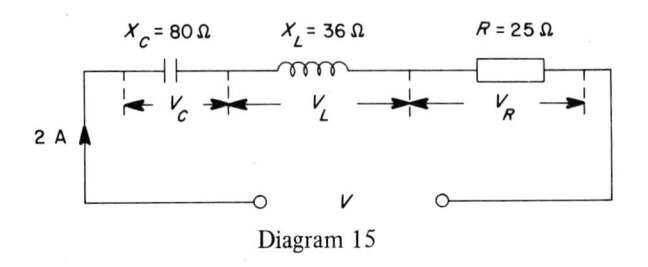

Diagram 15

Method 1, by phasors

$$V_C = I \times X_C = 2 \times 80 = 160 \text{ V}$$
$$V_L = I \times X_L = 2 \times 36 = 72 \text{ V}$$
$$V_R = I \times R = 2 \times 25 = 50 \text{ V}$$

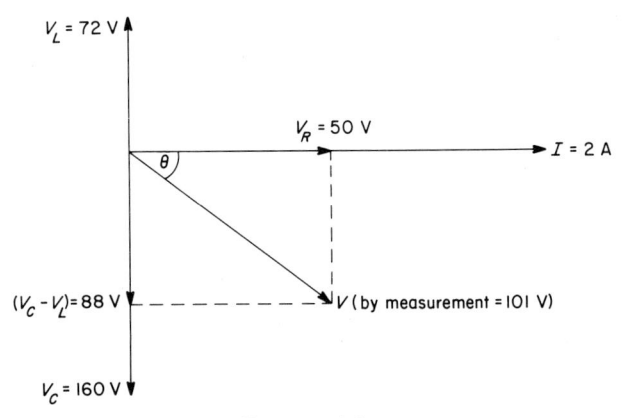

Diagram 16

From *Diagram 16*,

$$V = 101 \text{ V}$$

By measurement

$$\theta = 60.3°$$

From cosine tables,
$$\cos\theta = 0.495$$
$$\text{P.F.} = 0.495 \text{ leading}$$

Method 2, using the theorem of Pythagoras (impedance)

$$Z = \frac{V}{I}$$

But
$$Z = \sqrt{R^2 + (X_C - X_L)^2}$$
$$= \sqrt{25^2 + (80 - 36)^2}$$
$$= \sqrt{25^2 + 44^2}$$
$$= \sqrt{2561}$$
$$= 50.6\,\Omega$$
$$\therefore 50.6 = \frac{V}{2}$$
$$\therefore V = 2 \times 50.6$$
$$= 101.2 \text{ V}$$
$$\text{P.F.} = \cos\theta = \frac{R}{Z} = \frac{25}{50.6}$$
$$= 0.49 \text{ leading}$$

Method 3, using the theorem of Pythagoras (voltage)

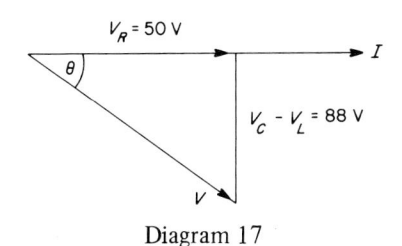

$V_R = 50$ V

$V_C - V_L = 88$ V

Diagram 17

16   Alternating-current theory

From *Diagram 17*,

$$V = \sqrt{V_R^2 + (V_C - V_L)^2}$$

$$= \sqrt{50^2 + 88^2}$$

$$= \sqrt{10244}$$

$$V = 101.2$$

$$\text{P.F.} = \cos\theta = \frac{\text{base}}{\text{hypotenuse}}$$

$$= \frac{50}{101.2}$$

$$= 0.49 \text{ leading}$$

Method 4, using trigonometry

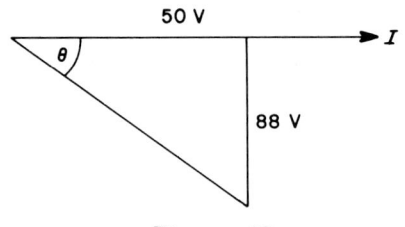

Diagram 18

From *Diagram 18*,

$$\tan\theta = \frac{\text{perpendicular}}{\text{base}}$$

$$= \frac{88}{50}$$

$$= 1.76$$

From tangent tables,

$$\theta = 60.4°$$

$$\cos\theta = \frac{\text{base}}{\text{hypotenuse}} = \frac{50}{V}$$

$$\therefore V = \frac{50}{\cos\theta}$$

From cosine tables,

$$\cos 60.4° = 0.49 \text{ (P.F.)}$$

$$\therefore V = \frac{50}{0.49}$$

$$= 101.2 \text{ V}$$

$$\text{P.F.} = \cos \theta = 0.49 \text{ leading}$$

R and L in series, in parallel with C

This circuit is typical of P.F. correction.

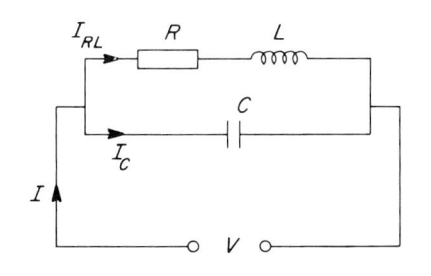

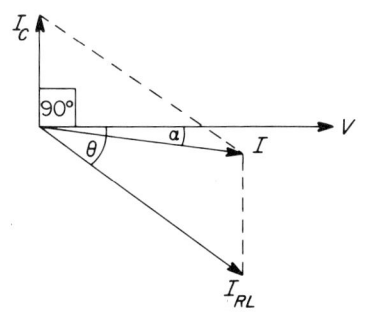

Circuit diagram   Phasor diagram

Diagram 19

From *Diagram 19*,

$$\text{Top branch:} \quad I_{RL} = \frac{V}{Z}$$

$$\cos \theta = \frac{R}{Z}$$

$$\text{Bottom branch:} \quad I_C = \frac{V}{X_C}$$

Example 2.3

A 240 V, 50 Hz single-phase motor takes a current of 8 A at a power factor of 0.65 lagging. Determine the value of capacitor required to improve the P.F. to 0.92 lagging. What is the value of the new supply current?

18 Alternating-current theory

Method 1, by phasors (Diagram 20)

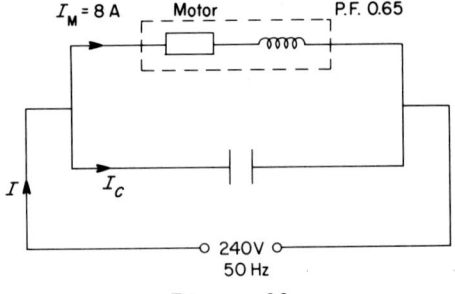

Diagram 20

$$\text{Old P.F.} = \cos\theta = 0.65$$

From cosine tables,

$$\theta = 49.5°$$

$$\text{New P.F.} = \cos\alpha = 0.92$$

From cosine tables

$$\alpha = 23°$$

The phasor diagram is now drawn to scale (*Diagram 21*). By measurement,

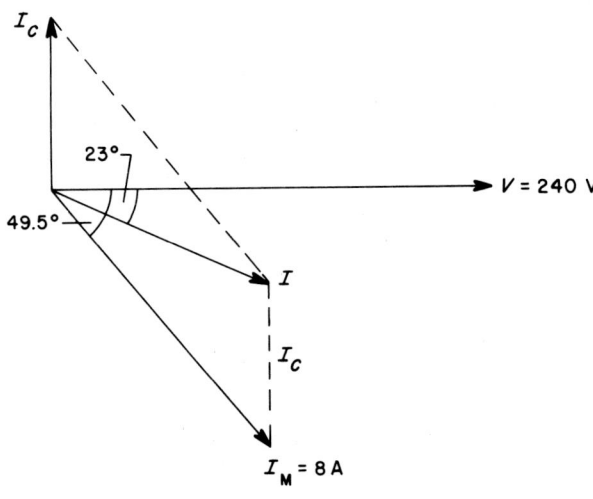

$$I_C = 3.9 \text{ A}$$

Diagram 21

and
$$I = 5.6 \text{ A}$$

To find the capacitance required,
$$I_C = \frac{V}{X_C}$$
$$\therefore X_C = \frac{V}{I_C}$$
$$= \frac{240}{3.9}$$
$$= 61.54 \, \Omega$$

But
$$X_C = \frac{1}{2\pi f C}$$
$$\therefore C = \frac{1}{2\pi f X_C}$$
$$= \frac{1}{2\pi \times 50 \times 61.54}$$
$$= 51.7 \, \mu\text{F}$$

Method 2, by trigonometry

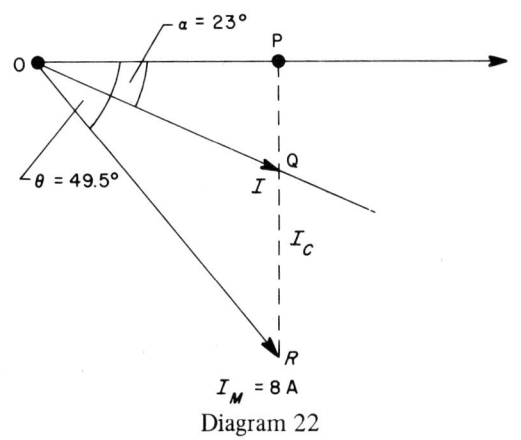

Diagram 22

20 Alternating-current theory

In this method, I_C is found by calculating lengths PR and PQ in *Diagram 22* and subtracting (PR − PQ = QR = I_C).

Triangle OPR:

$$\cos \theta = \frac{\text{base}}{\text{hypotenuse}} = \frac{\text{OP}}{\text{OR}}$$

$$\therefore \text{OP} = \text{OR} \times \cos \theta$$

$$= 8 \times 0.65$$

$$\text{OP} = 5.2 \text{ A}$$

This is called the active or horizontal component of I_M.

$$\sin \theta = \frac{\text{perpendicular}}{\text{hypotenuse}} = \frac{\text{PR}}{\text{OR}}$$

$$\therefore \text{PR} = \text{OR} \times \sin \theta$$

$$= 8 \times 0.76$$

$$\text{PR} = 6.08 \text{ A}$$

This is called the reactive or vertical component of I_M.

Triangle OPQ:

$$\tan \alpha = \frac{\text{perpendicular}}{\text{base}} = \frac{\text{PQ}}{\text{OP}}$$

$$\therefore \text{PQ} = \text{OP} \times \tan \alpha$$

$$= 5.2 \times 0.424$$

$$\text{PQ} = 2.2 \text{ A}$$

$$\therefore \text{QR} = I_C = (\text{PR} - \text{PQ})$$

$$= 6.08 - 2.2$$

$$I_C = 3.88 \text{ A}$$

Calculation of C is as for method 1. To find current I:

$$\cos \alpha = \frac{\text{base}}{\text{hypotenuse}} = \frac{OP}{OQ}$$

$$\therefore OQ = I = \frac{OP}{\cos \alpha}$$

$$= \frac{5.2}{0.92}$$

$$I = 5.65 \text{ A}$$

Example 2.4

Two 240 V fluorescent lamps are arranged to overcome the stroboscopic effect. One unit takes 0.8 A at 0.45 P.F. leading, the other takes 0.7 A at 0.5 P.F. lagging (*Diagram 23*). Determine the total current drawn and the overall P.F.

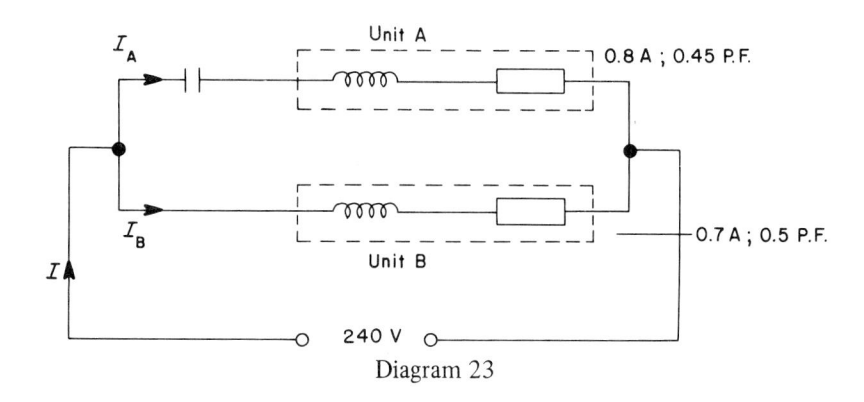

Diagram 23

Method 1, by phasors (Diagram 24)

$$\text{Unit A: } \cos \theta = 0.45$$

$$\therefore \theta = 63.25°$$

$$\text{Unit B: } \cos \alpha = 0.5$$

$$\therefore \alpha = 60°$$

22 Alternating-current theory

Diagram 24

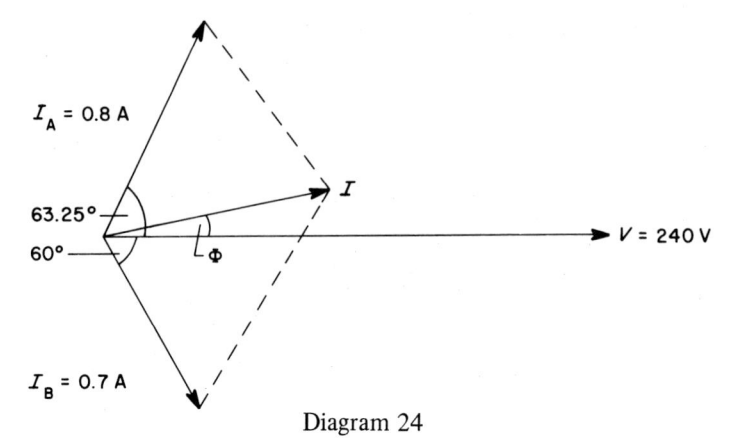

By measurement,

$$I = 0.71 \text{ A}$$

and

$$\Phi = 8.5°$$

$$\therefore \cos \Phi = \text{P.F.} = 0.989 \text{ leading}$$

Method 2, by trigonometry

In this method, the active and reactive components of currents I_A and I_B are found (*Diagram 25*).

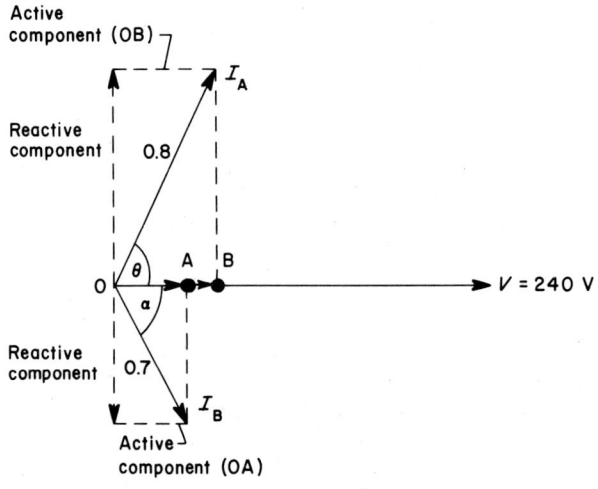

Diagram 25

Active or horizontal component of I_A = OB = $I_A \times \cos\theta$

$$\therefore \text{OB} = 0.8 \times \cos 63.25°$$
$$= 0.8 \times 0.45$$
$$= 0.36 \text{ A}$$

Active component of I_B = OA = $I_B \times \cos\alpha$

$$\therefore \text{OA} = 0.7 \times \cos 60°$$
$$= 0.7 \times 0.5$$
$$= 0.35 \text{ A}$$

Since all the active components are in phase they may be added. Hence

Total of active components = 0.36 + 0.35 = 0.71 A

Reactive or vertical component of I_A = $I_A \times \sin\theta$

$$= 0.8 \times \sin 63.25°$$
$$= 0.8 \times 0.893$$
$$= 0.714 \text{ A}$$

Reactive component of I_B = $I_B \times \sin\alpha$

$$= 0.7 \times \sin 60°$$
$$= 0.7 \times 0.866$$
$$= 0.606 \text{ A}$$

Since the two reactive components are opposite in phase they must be subtracted,

$$\therefore \text{Total of reactive components} = 0.714 - 0.606$$
$$= 0.108 \text{ A}$$

The resultant of the total active and reactive components will be the total current taken (*Diagram 26*).

24 Alternating-current theory

Diagram 26

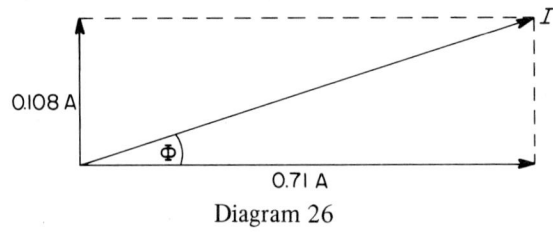

$$\tan \Phi = \frac{\text{perpendicular}}{\text{base}}$$

$$= \frac{0.108}{0.71}$$

$$\tan \Phi = 0.152$$

$$\therefore \Phi = 8.64°$$

$$\therefore \text{P.F.} = \cos \Phi = 0.988 \text{ leading}$$

and

$$\cos \Phi = \frac{\text{base}}{\text{hypotenuse}} = \frac{0.71}{I}$$

$$\therefore I = \frac{0.71}{\cos \Phi} = \frac{0.71}{0.988}$$

$$I = 0.718 \text{ A}$$

Power in a.c. circuits

The phasor diagram of voltages in an a.c. series circuit can be used to give a power triangle (*Diagram 27*).

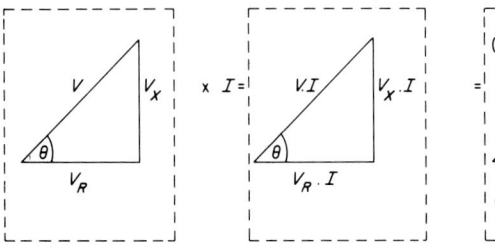

Diagram 27

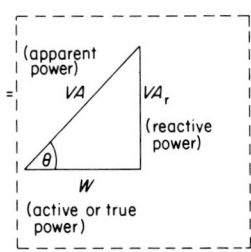

Alternating-current theory 25

$$\cos \theta = \frac{W}{VA} = \text{P.F.}$$

Power may be added by phasor diagram or calculated by trigonometry.

Example 2.5

The following loads are connected to a factory supply: 5 kVA at 0.75 P.F. lagging; 8 kW at a P.F. of unity; 6.8 kVA at 0.6 P.F. lagging. Determine the total load taken from the supply and the overall P.F.

Method 1, by phasor diagram (Diagram 28)

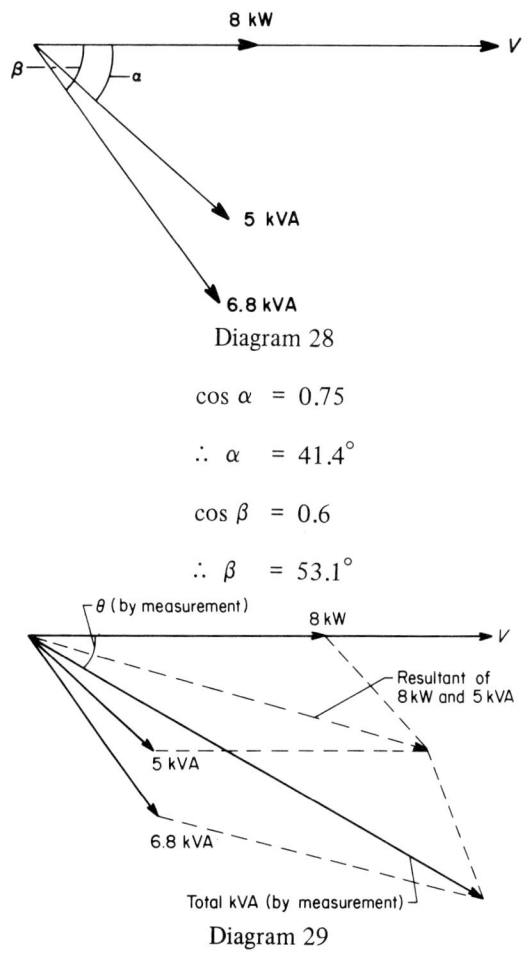

Diagram 28

$$\cos \alpha = 0.75$$
$$\therefore \alpha = 41.4°$$
$$\cos \beta = 0.6$$
$$\therefore \beta = 53.1°$$

Diagram 29

26 Alternating-current theory

From *Diagram 29*,

$$\text{Total kVA} = 18.1 \text{ kVA}$$

$$\theta = 29°$$

$$\therefore \cos \theta = \text{P.F.} = 0.874$$

Method 2, by trigonometry (Diagram 30)

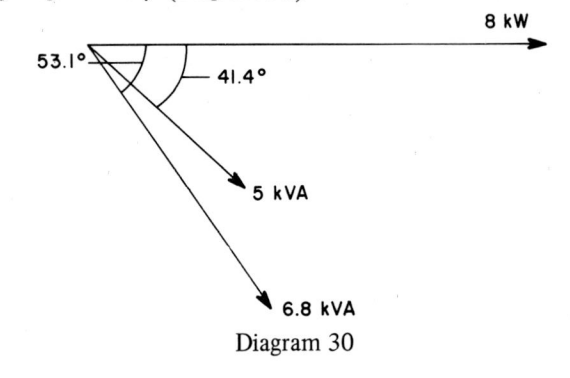

Diagram 30

Active component of 8 kW = 8 kW

Active component of 5 kVA = 5 × cos 41.4°
= 5 × 0.75
= 3.75 kW

Active component of 6.8 kVA = 6.8 × cos 53.1°
= 6.8 × 0.6
= 4.08 kW

Total of active components = 8 + 3.75 + 4.08
= 15.83 kW

Reactive component of 8 kW = 0

Reactive component of 5 kVA = 5 × sin 41.4°
= 5 × 0.66
= 3.3 kVA$_r$

Reactive component of 6.8 kVA = 6.8 × sin 53.1°
= 6.8 × 0.8
= 5.44 kVA$_r$

As both 5 kVA and 6.8 kVA have lagging power factors, their reactive components are added (*Diagram 31*).

∴ Total reactive component = 0 + 3.3 + 5.44
= 8.74 kVA$_r$

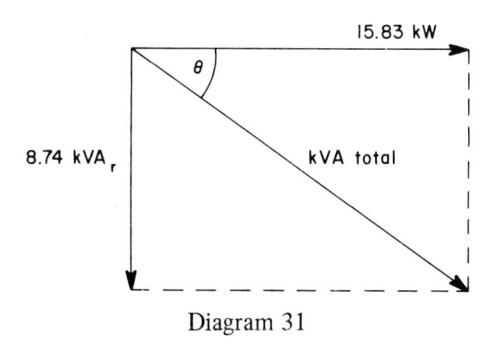

Diagram 31

$$\tan \theta = \frac{8.74}{15.83} = 0.55$$

∴ $\theta = 28.9°$

∴ P.F. = $\cos \theta$ = 0.875

Also

$$\cos \theta = \frac{15.83}{kVA}$$

∴ $kVA = \frac{15.83}{\cos \theta}$

$= \frac{15.83}{0.875}$

= 18 kVA at 0.875 P.F. lagging

28 Alternating-current theory

Balanced three-phase a.c. systems (*Diagram 32*)

Star connection   Delta connection

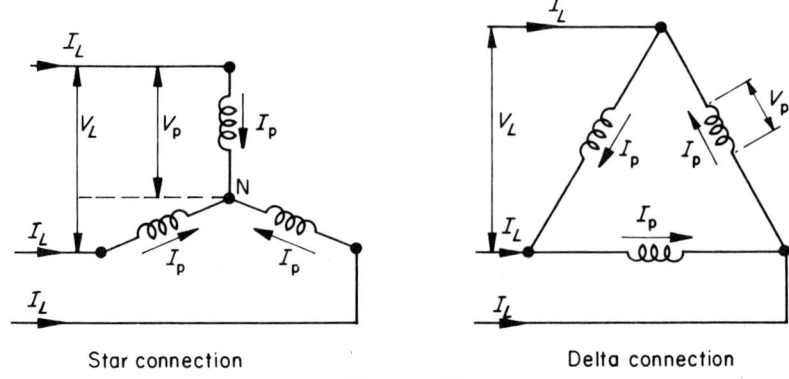

Diagram 32

$$I_L = I_p \qquad\qquad I_L = \sqrt{3} \times I_p$$
$$V_L = \sqrt{3} \times V_p \qquad\qquad V_L = V_p$$

Power in a star-connected system

Since we are considering balanced systems, the total power is three times the power in one phase, and as

$$\text{P.F.} = \frac{W}{VA}$$

then

$$W = VA \times \text{P.F.}$$

$$\therefore W = V_p \times I_p \times \text{P.F.}$$

But

$$V_p = \frac{V_L}{\sqrt{3}} \quad \text{and} \quad I_p = I_L$$

$$\therefore W = \frac{V_L}{\sqrt{3}} \times I_L \times \text{P.F.}$$

$$\therefore \text{Total power } P = 3 \times \frac{V_L}{\sqrt{3}} \times I_L \times \text{P.F.}$$

$$= \sqrt{3}\, V_L \times I_L \times \text{P.F.}$$

Delta-connected system

$$\text{Again} \quad W = VA \times P.F.$$

$$\therefore W = V_p \times I_p \times P.F.$$

But

$$V_p = V_L \text{ and } I_p = \frac{I_L}{\sqrt{3}}$$

$$\therefore W = V_L \times \frac{I_L}{\sqrt{3}} \times P.F.$$

$$\therefore \text{Total power } P = 3 \times V_L \times \frac{I_L}{\sqrt{3}} \times P.F.$$

$$P = \sqrt{3}\, V_L \times I_L \times P.F.$$

which is the same as for a star connection.

Hence for either star or delta connections the total power in watts is given by

$$P \text{ (watts)} = \sqrt{3}\, V_L \times I_L \times P.F.$$

Since

$$P.F. = \frac{W}{VA}$$

then

$$VA = \frac{W}{P.F.}$$

$$\therefore \text{Three-phase VA} = \sqrt{3}\, V_L \times I_L$$

$$\therefore \text{Line current} = \frac{VA}{\sqrt{3}\, V_L}$$

Example 2.6

A 15 kW, 415 V, balanced three-phase delta-connected load has a power factor of 0.8 lagging. Calculate the line and phase currents.

Alternating-current theory

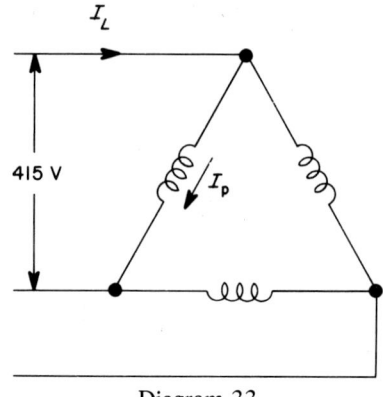

Diagram 33

From *Diagram 33*,

$$P = \sqrt{3}\, V_L \times I_L \times \text{P.F.}$$

$$\therefore I_L = \frac{P}{\sqrt{3}\, V_L \times \text{P.F.}}$$

$$= \frac{15\,000}{\sqrt{3} \times 415 \times 0.8}$$

$$I_L = 26\,\text{A}$$

For delta connections

$$I_L = \sqrt{3}\, I_p$$

$$\therefore I_p = \frac{I_L}{\sqrt{3}}$$

$$= \frac{26}{\sqrt{3}}$$

$$I_p = 15\,\text{A}$$

Example 2.7

Three identical loads each having a resistance of 10 Ω and an inductive reactance of 20 Ω are connected first in star and then in delta across a 415 V, 50 Hz three-phase supply. Calculate the line and phase currents in each case.

$$Z \text{ for each load} = \sqrt{R^2 + X_L^2}$$
$$= \sqrt{10^2 + 20^2}$$
$$= \sqrt{500}$$
$$= 22.36 \, \Omega$$

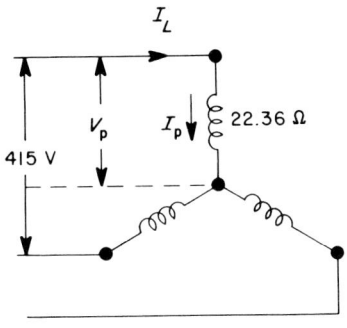

Diagram 34

For star connection (*Diagram 34*),

$$V_p = \frac{V_L}{\sqrt{3}} = 240 \text{ V}$$

$$I_L = I_p = \frac{V_L}{Z}$$

$$= \frac{240}{22.36}$$

$$I_L = I_p = 10.73 \text{ A}$$

For delta connection (*Diagram 35*),

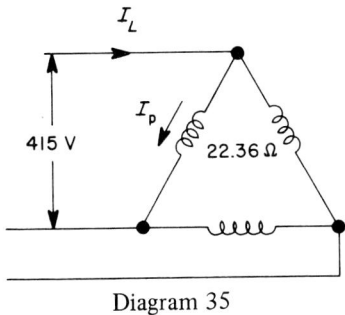

Diagram 35

32 Alternating-current theory

$$V_L = V_p = 415\,\text{V}$$

$$\therefore I_p = \frac{V_p}{22.36}$$

$$= \frac{415}{22.36}$$

$$I_p = 18.56\,\text{V}$$

and

$$I_L = \sqrt{3}\,I_p$$

$$= \sqrt{3} \times 18.56$$

$$I_L = 32.15\,\text{A}$$

Measurement of power in a.c. circuits

Single-phase circuit

Diagram 36 shows the instruments required to determine the power in watts and volt-amperes, and the power factor of a single-phase circuit.

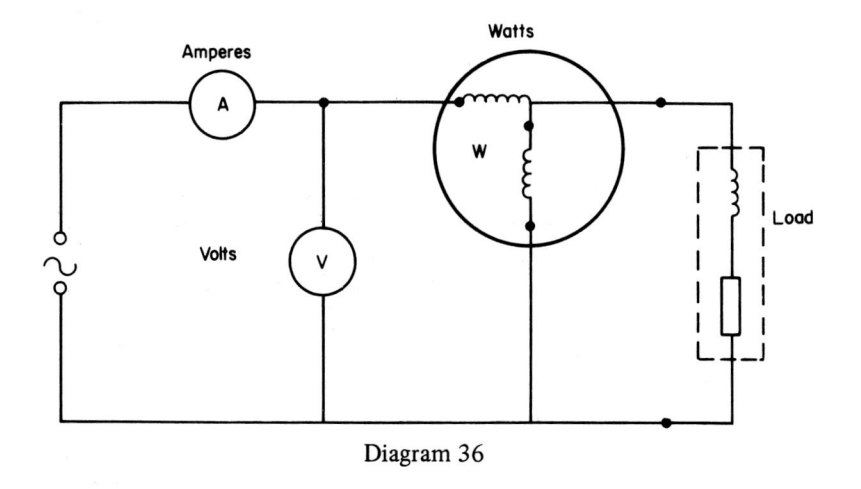

Diagram 36

Alternating-current theory 33

Example 2.8

The following values were recorded from a circuit similar to that in *Diagram 36*.

Ammeter — 8 A
Voltmeter — 240 V
Wattmeter — 1.152 kW

Calculate the kVA and the power factor of the load.

$$kVA = \frac{VA}{1000}$$

$$= \frac{8 \times 240}{1000}$$

$$= 1.92 \, kVA$$

$$P.F. = \frac{kW}{kVA}$$

$$= \frac{1.152}{1.92}$$

$$= 0.6$$

Three-phase (balanced) four-wire circuit

In this case it is necessary to measure the power only in one phase. The total power will be three times this value. Also, the P.F. for one phase is the overall P.F.

Example 2.9

It is required to measure the power factor of a three-phase star-connected balanced inductive load. Show how the necessary instruments would be arranged, a single voltmeter being used to measure the voltage across each phase.
 If the readings obtained were 20 A, 240 V and 3840 W, calculate the total power in kilowatts and the power factor.

From *Diagram 37*,

$$P.F. = \frac{W}{VA}$$

34 *Alternating-current theory*

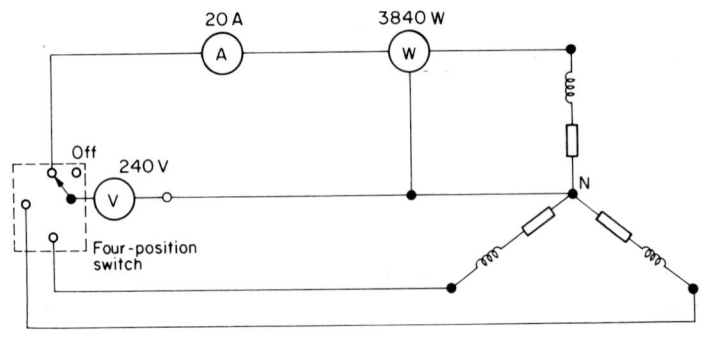

Diagram 37

$$= \frac{3840}{240 \times 20} = 0.8 \text{ lagging}$$

Total power = 3 × 3840
= 11 520
= 11.52 kW

Check:

$$V_L = V_p\sqrt{3}$$

$$V_L = 240 \times \sqrt{3}$$

$$P = \sqrt{3}\ V_L \times I_L \times \text{P.F.}$$

$$= \sqrt{3} \times \sqrt{3} \times 240 \times 20 \times 0.8$$

$$= 11.52 \text{ kW}$$

Questions on Chapter 2

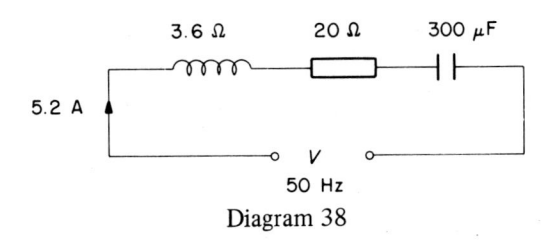

Diagram 38

Alternating-current theory 35

1. Determine the value of the voltage and the power factor in the circuit shown in *Diagram 38*.
2. A 240 V, 50 Hz single-phase motor takes 6 A at 0.56 P.F. lagging. Determine the value of capacitor required to improve the P.F. to unity.
3. A 240 V, 50 Hz fluorescent lamp unit takes a current of 0.6 A at a P.F. of 0.45 lagging. Calculate the capacitance required to correct the P.F. to 0.92 lagging.
4. Two 240 V, 50 Hz fluorescent lamp units A and B are arranged so as to overcome stroboscopic effects. Unit A takes a current of 0.75 A at a P.F. of 0.51 leading, unit B takes 0.9 A at a P.F. of 0.43 lagging. Calculate the total current taken from the supply and the overall P.F.
5. Two 240 V, 50 Hz single-phase motors A and B are connected in parallel. Motor A takes a current of 8.6 A at 0.75 P.F. lagging and the total current taken from the supply is 16 A at 0.6 lagging. Calculate the current and P.F. of motor B.
6. A consumer has the following loads connected to his supply: 3 kVA at 0.8 lagging; 4 kW at a P.F. of unity; and 5 kVA at 0.5 lagging. Calculate the total load in kVA and the overall P.F.
7. A voltmeter, ammeter and wattmeter are arranged to measure the power in a single-phase circuit. Show how these instruments would be connected and calculate the circuit P.F. if the readings were 16 A, 240 V and 3600 W.
8. A three-phase star-connected load is supplied from the delta-connected secondary of a transformer. If the transformer line voltage is 190.5 V and the load phase current is 10 A, calculate the transformer phase current and the load phase voltage.
9. A 20 kW, 415 V three-phase star-connected load takes a line current of 34.8 A. Calculate the P.F. of the load.
10. A wattmeter, voltmeter and ammeter are arranged in a three-phase four-wire balanced system. Show how they would be connected ready to determine the total circuit power in watts and the P.F. Readings obtained were 2000 W, 240 V and 10 A. Calculate the circuit power in kW and kVA.

CHAPTER 3

Machines

Electrical machines play an important part in the modern domestic and industrial environment, and should therefore be of interest to the electrician.
    Machines may be divided into two distinct kinds: those using direct current and those using alternating current. Each of these categories is further divided into different types.

Direct-current machines

Principle of operation

If a current is passed through a conductor situated at right angles to a magnetic field, a force will be exerted on that conductor (*Diagrams 39* and *40*).

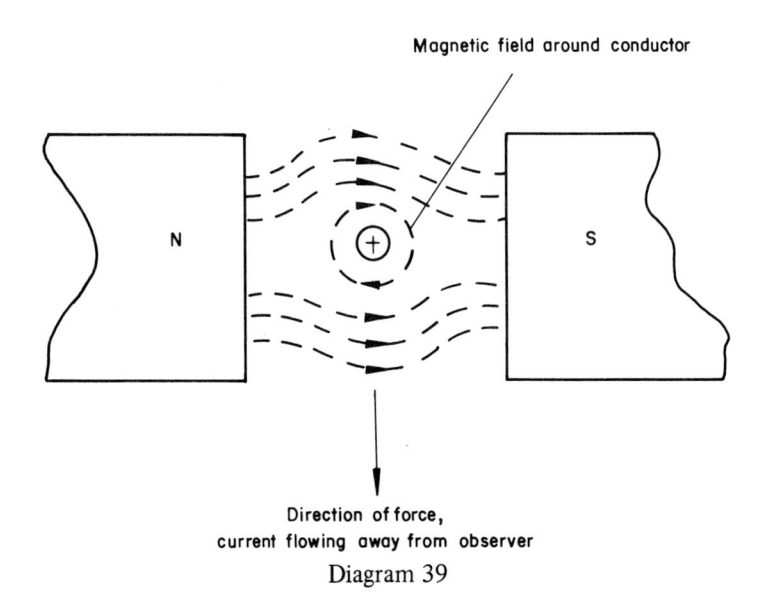

Magnetic field around conductor

Direction of force,
current flowing away from observer

Diagram 39

Direction of force

Current flowing towards observer

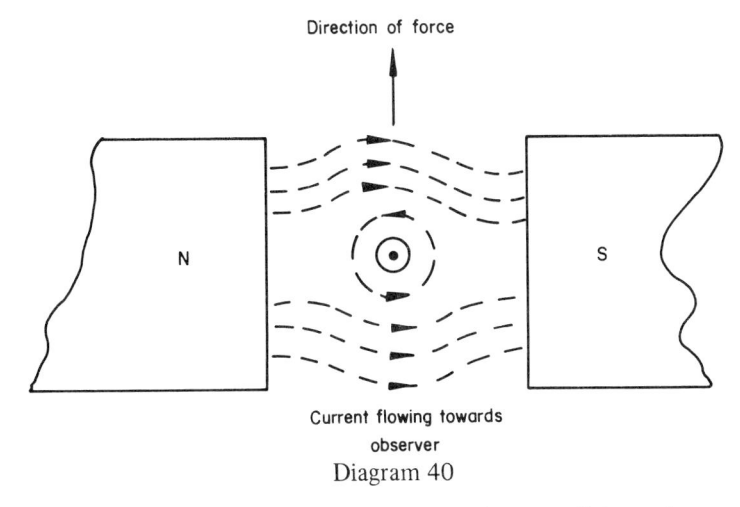

Diagram 40

The strength of the force acting on the conductor will depend on:

(a) the strength or flux density of the main magnetic field;
(b) the magnitude of the current flowing in the conductor; and
(c) the length of the conductor in the magnetic field.

Therefore
$$F = B \times l \times I \quad \text{newtons}$$

where F is the force in newtons, B is the flux density in teslas, l is the length of the conductor in metres and I is the current flowing, in amperes.

Example 3.1

Calculate the current flowing in a conductor 6 cm long if the density of the magnetic field in which the conductor is situated is 30 T and the force on the conductor is 18 N.

$$F = B \times l \times I$$

$$\therefore I = \frac{F}{B \times l}$$

$$= \frac{18}{30 \times 6 \times 10^{-2}}$$

$$= 10 \text{ A}$$

38 Machines

Example 3.2

A conductor 10 cm long is situated at right angles to a magnetic field of cross-sectional area 40 cm² and flux 80 mWb. If the current flowing is 25 A, calculate the force on the conductor.

$$F = B \times I \times l$$

and

$$B = \frac{\Phi}{a}$$

$$\therefore B = \frac{80 \times 10^{-3}}{40 \times 10^{-4}}$$

$$= 20 \, \text{T}$$

$$\therefore F = 20 \times 10 \times 10^{-2} \times 25$$

$$= 50 \, \text{N}$$

Simple single-loop motor

The effect of the force on a conductor in a magnetic field may be used to cause the rotation of a motor armature. *Diagram 41* illustrates a simple single-loop motor.

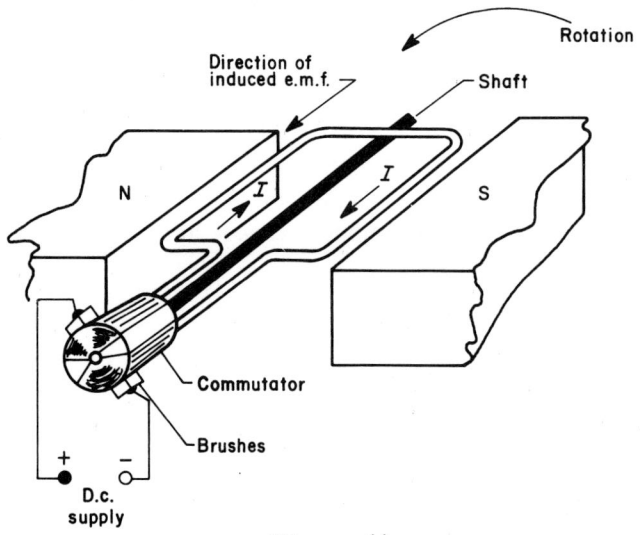

Diagram 41

In practice, the d.c. motor comprises an armature of many loops revolving between electromagnetic poles. Both the armature and the field are supplied from the same source. The commutator has, of course, many segments to which the ends of the armature coils are connected, and the armature core is laminated to reduce eddy currents.

Back e.m.f.: symbol, E; unit, volt (V)

It is interesting to note that as the armature revolves, its coils cut across the field flux, but we know that if a conductor cuts across lines of force, an e.m.f. is induced in that conductor. This is the principle of the generator, as dealt with in Volume 1. So, applying Fleming's right-hand rule to *Diagram 41*, we see that the induced e.m.f. is opposing the supply. This induced e.m.f. is called the 'back e.m.f.'. If the back e.m.f. were of the same magnitude as the supply voltage, no current would flow and the motor would not work. As current must flow in the armature to produce rotation, and as the armature circuit has resistance, then there must be a volt drop in the armature circuit. This volt drop is the product of the armature current (I_a) and the armature circuit resistance (R_a).

$$\text{Armature volt drop} = I_a \times R_a$$

It is this volt drop that is the difference between the supply voltage and the back e.m.f. Hence

$$E = V - (I_a \times R_a)$$

We also know (from Volume 1) that induced e.m.f. is dependent on the flux density (B), the speed of cutting the flux (v) and the length of the conductor (l).

$$E = B \times l \times v$$

But

$$B = \frac{\Phi}{a} \left(\frac{\text{flux}}{\text{area}} \right)$$

$$\therefore E = \frac{\Phi}{a} \times l \times v$$

Both l and a for a given conductor will be constant and v is replaced by n (revs/second) as this represents angular or rotational speed.

$$\therefore E \propto n\Phi$$

40 Machines

So, if the speed is changed from n_1 to n_2 and the flux from Φ_1 to Φ_2, then the e.m.f. will change from E_1 to E_2.

$$\therefore \frac{E_1}{E_2} = \frac{n_1 \Phi_1}{n_2 \Phi_2}$$

Torque: symbol, T; unit, newton-metre (Nm)

$$\text{Work} = \text{force} \times \text{distance}$$

$$\therefore \text{Turning work or torque} = \underset{(F)}{\text{force}} \times \underset{(r)}{\text{radius}}$$

We also know that

$$\text{Force} = B \times l \times I$$

$$\therefore \text{Torque } T = B \times l \times I_a \times r$$

$$\therefore T = \frac{\Phi}{a} \times l \times I_a \times r$$

Once again, for a given machine, a, l and r will all be constant.

$$\therefore T \propto \Phi \times I_a$$

Also, mechanical output power in watts is given by

$$P = 2\pi n T$$

where P is the output power in watts, n is the speed in revs/second and T is the torque in newton-metres.

If we multiply E ($= V - I_a R_a$) by I_a, we get

$$E I_a = V I_a - I_a^2 R_a$$

$V I_a$ is the power supplied to the armature and $I_a^2 R_a$ is the power loss in the armature, therefore $E I_a$ must be the armature power output. Hence

$$E I_a = P$$

$$\therefore E I_a = 2\pi n T$$

$$\therefore T = \frac{E I_a}{2\pi n} \text{ newton-metres}$$

It is clear then that torque is directly proportional to armature current and inversely proportional to speed, i.e. if the mechanical load is lessened, the torque required is less, the armature current decreases and the motor speeds up.

Example 3.3

A 300 V d.c. motor runs at 15 revs/second and takes an armature current of 30 A. If the armature resistance is 0.5 Ω, calculate first the back e.m.f. and secondly the torque.

$$E = V - I_a R_a$$

$$= 300 - (30 \times 0.5)$$

$$= 300 - 15$$

$$E = 285 \text{ V}$$

$$T = \frac{E I_a}{2\pi n}$$

$$= \frac{285 \times 30}{2\pi \times 15}$$

$$T = 90.72 \text{ Nm}$$

Series motor

The series type of d.c. motor has its field windings and armature connected in series across the supply (*Diagram 42*). It will be seen from this diagram that the armature current I_a also supplies the field. Therefore, when I_a is large (on starting, for example), the magnetic field will be strong, and the torque will be

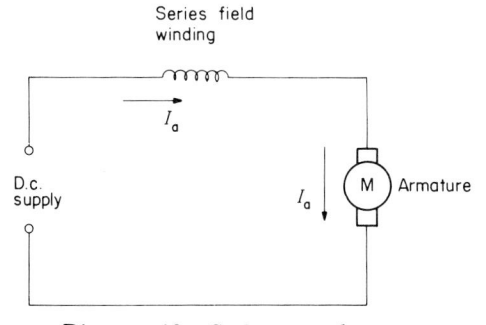

Diagram 42   Series-wound motor

42 Machines

high. As the machine accelerates, the torque, armature current and field strength will all decrease. This type of motor should never be coupled to its load by means of a belt, since if the belt breaks the required torque from the armature will be removed, the armature and field current will fall, lightening the magnetic field, and the motor will increase in speed until it disintegrates. *Diagram 43* shows the graphs of speed and torque to a base of load current.

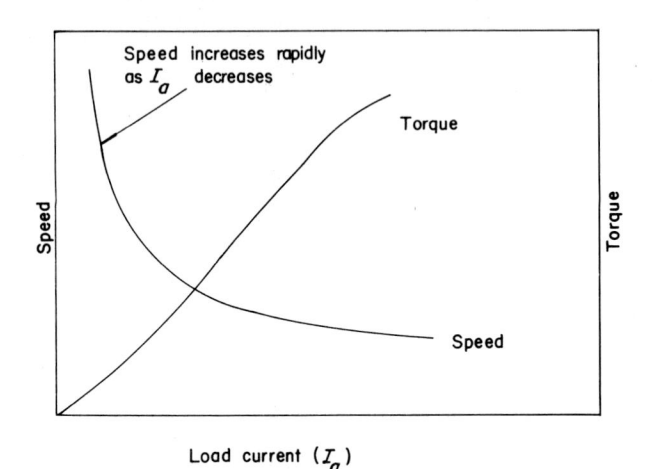

Diagram 43   Load characteristics of a series motor

Speed control

The most effective way of controlling the speed of a d.c. motor is to vary the strength of the magnetic field. On a series machine this may be achieved by diverting some of the current through a variable resistor (*Diagram 44*).

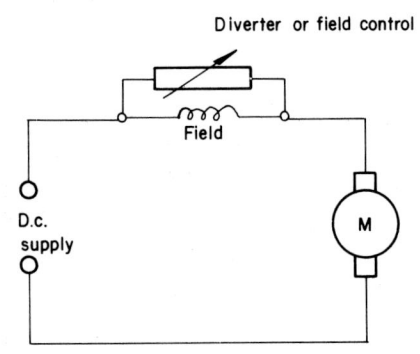

Diagram 44   Speed control of a series motor

Starting

A series motor is started by placing a variable resistor in series with the armature circuit (*Diagram 45*).

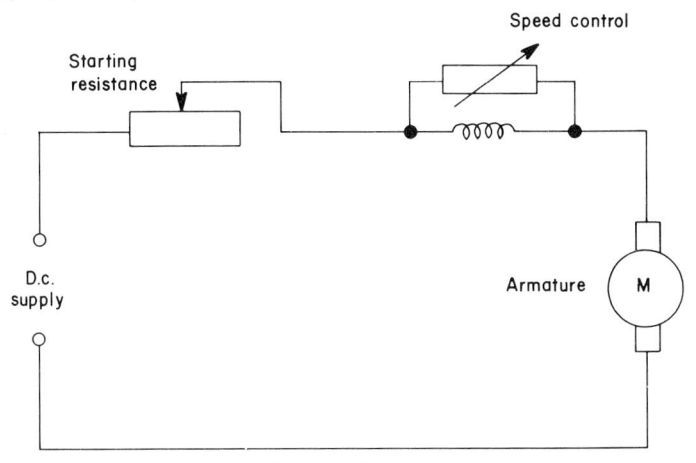

Diagram 45 Starting a series motor

Applications

The series motor is best used where heavy masses need to be accelerated from rest, such as in cranes and lifts.

Example 3.4

A 200 V series motor has a field winding resistance of 0.1 Ω and an armature resistance of 0.3 Ω. If the current taken at 5 revs/second is 30 A, calculate the torque on the armature.

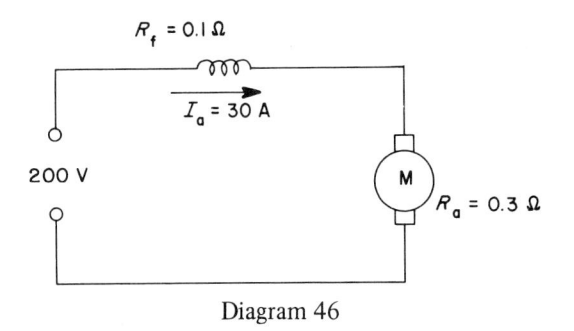

Diagram 46

The total armature circuit resistance is $R_a + R_f$ (*Diagram 46*) as the two resistances are in series.

$$E = V - I_a R_a$$
$$= 200 - 30(0.1 + 0.3)$$
$$= 200 - 30 \times 0.4$$
$$= 200 - 12$$
$$= 188 \text{ V}$$

$$\text{Torque } T = \frac{EI_a}{2\pi n}$$
$$= \frac{188 \times 30}{2\pi \times 5}$$
$$= 180 \text{ Nm}$$

Shunt motor

In the case of a shunt motor, the motor winding is in parallel with the armature (*Diagram 47*). It will be seen here that the supply current $I = I_a + I_f$. Unlike the series motor, if the load is removed from the motor, only the armature current will decrease, the field remaining at the same strength. The motor will therefore not continue to speed up to destruction. *Diagram 48* shows the graphs of speed and torque to a base of load current.

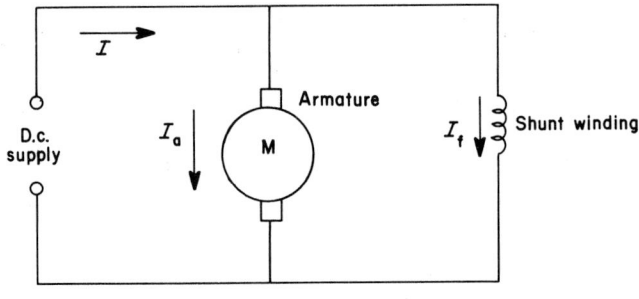

Diagram 47 Shunt-wound motor

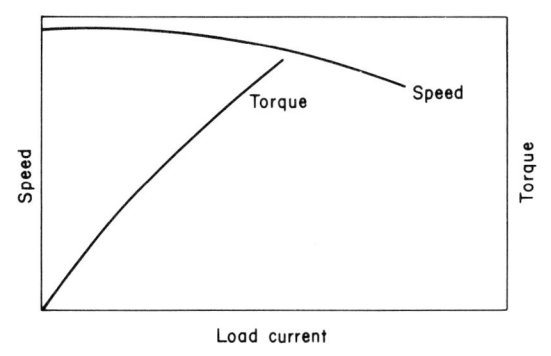

Diagram 48   Load characteristics of a shunt motor

Speed control

As in a series motor, speed control is best achieved by controlling the field strength, and in shunt motors a variable resistance is placed in series with the shunt winding (*Diagram 49*).

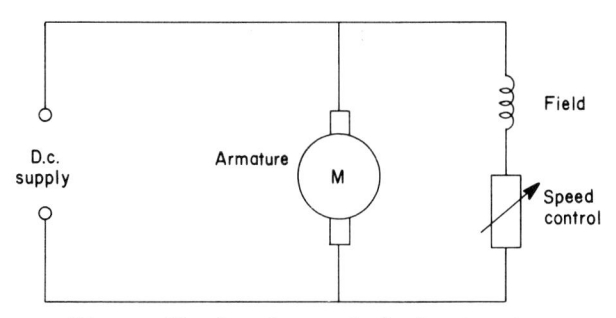

Diagram 49   Speed control of a shunt motor

Starting

Starting large d.c. shunt-wound motors is usually carried out using a d.c. face-plate starter (*Diagram 50*).

The face-plate starter comprises the following items:

(a) a series of resistances connected to brass studs;

(b) a spring-loaded handle which makes contact with two brass strips and also the brass studs;

(c) a no-volt release; and

(d) an overload release.

46 *Machines*

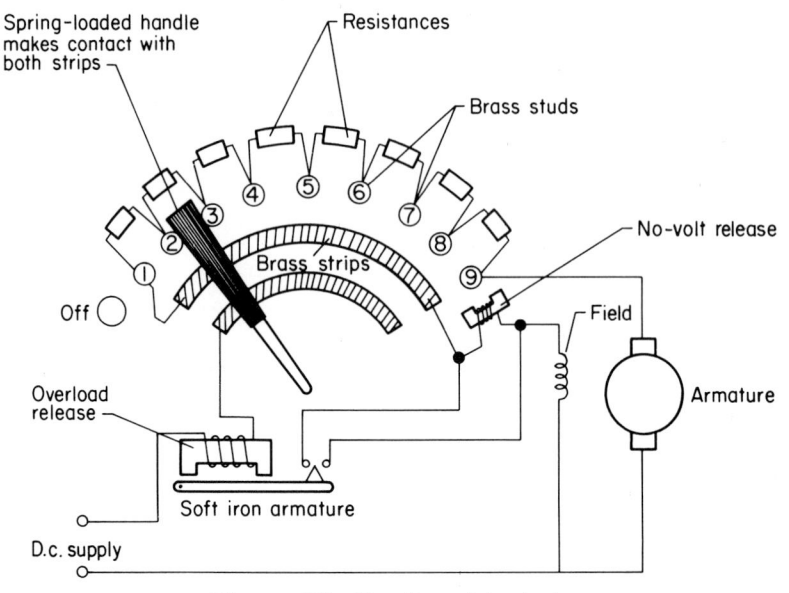

Diagram 50   D.c. face-plate starter

When the handle is located on the first stud, the field is supplied via the overload release, the top brass strip via the handle and the no-volt release (note that the field is continuously supplied in this way). The armature is supplied via the resistances.

As the handle is moved round, the resistance in the armature circuit is gradually cut out. On the final stud the handle is held in place by the no-volt release electromagnet. Should a failure in supply occur, the no-volt release will de-energize and the handle will spring back to the 'off' position. If a serious overload occurs, the overload release will energize sufficiently to attract its soft iron armature which will short out the no-volt release coil, and the handle will return to the 'off' position.

Note: The handle should be moved slowly from stud to stud.

Applications

As the speed of a shunt motor is almost constant over a wide range of loads, it is most suitable for small machine tools.

Example 3.5

A 400 V shunt-wound motor has a field winding resistance of 200 Ω and an armature resistance of 0.5 Ω. If the current taken from the supply is 22 A, calculate the back e.m.f.

Diagram 51

From *Diagram 51*,

$$I = I_a + I_f$$

$$\therefore I_a = I - I_f$$

$$I_f = \frac{V}{R_f} = \frac{400}{200} = 2 \text{ A}$$

$$\therefore I_a = 22 - 2 = 20 \text{ A}$$

$$E = V - I_a R_a$$

$$= 400 - (20 \times 0.5)$$

$$= 400 - 10$$

$$= 390 \text{ V}$$

Compound motor

A compound motor is a combination of a series and a shunt-wound motor (*Diagrams 52a* and *52b*).

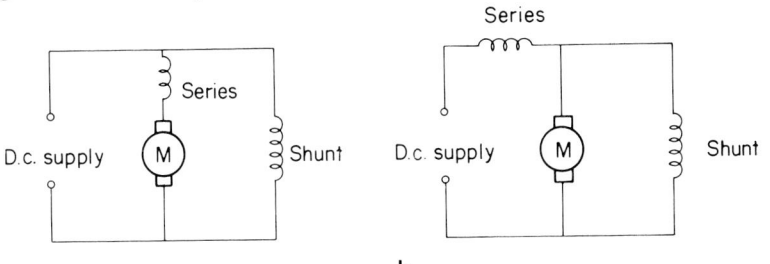

a b
Diagram 52 Compound-wound motor. (a) Long shunt; (b) short shunt

48 *Machines*

Diagram 53 shows the speed-torque characteristics for the compound motor.

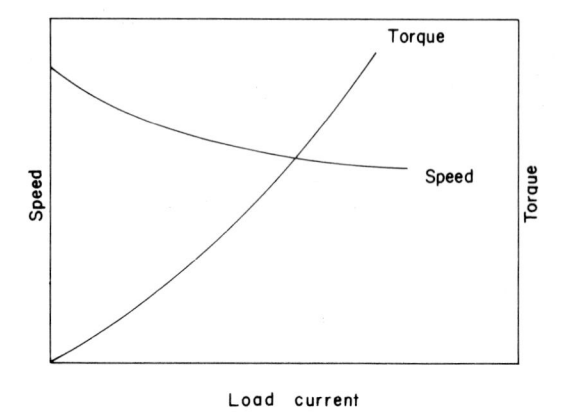

Diagram 53   Speed and torque characteristics of a compound motor

Speed control

Speed is usually controlled by variable resistors in the shunt field and armature circuit (*Diagram 54*).

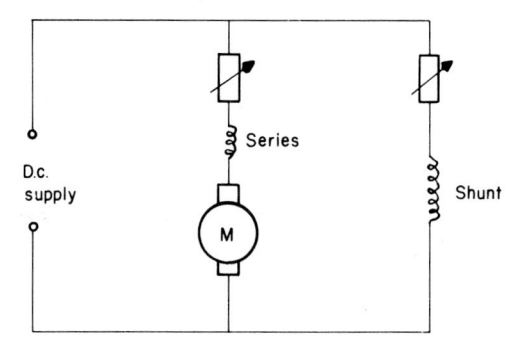

Diagram 54   Speed control of a typical compound motor (long shunt)

The series winding may be arranged such that it aids the shunt field (*cumulative* compound) or opposes it (*differential* compound).

Cumulatively compounded motors are similar in characteristics to series motors, while differentially compounded motors are similar to shunt motors (*Diagram 55*).

[Graph: Speed vs Load current, showing "Differential" curve rising and "Cumulative" curve falling]

Diagram 55

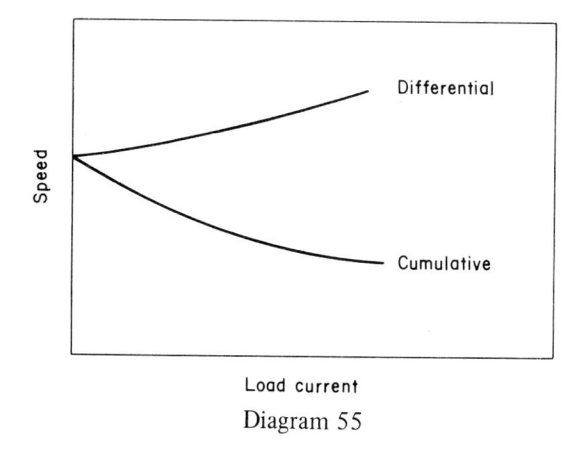

Starting

The d.c. face-plate starter is suitable for the compound motor.

Applications

These motors can be used for applications where a wide speed range is required. However, the differentially compounded type is rarely used as it tends to be unstable. Cumulatively compounded types are suitable for heavy machine tools.

Reversing d.c. motors

D.c. motors may be reversed in direction by altering either the polarity of the field or that of the armature. This is done by reversing the connection to the armature *or* the field winding.

D.c. generators

If by some means a d.c. motor is supplied with motive power it will act as a generator. Connection of the field windings is the same as for motors; that is, either series, shunt or compound.

E.m.f. generated: symbol, E; unit, volt (V)

When the armature is rotated in the field, an e.m.f. is induced or generated in the armature windings. When an external load is connected, current (I_a) will flow from the armature; this will cause a volt drop of $I_a R_a$ where R_a is the resistance of the armature circuit. Hence the voltage available at the load is less than the generated e.m.f.

$$\therefore E = V + I_a R_a$$

Example 3.6

A 240 V shunt-wound generator has a field resistance of 120 Ω and an armature resistance of 0.4 Ω. Calculate the generated e.m.f. when it is delivering 20 A to the load.

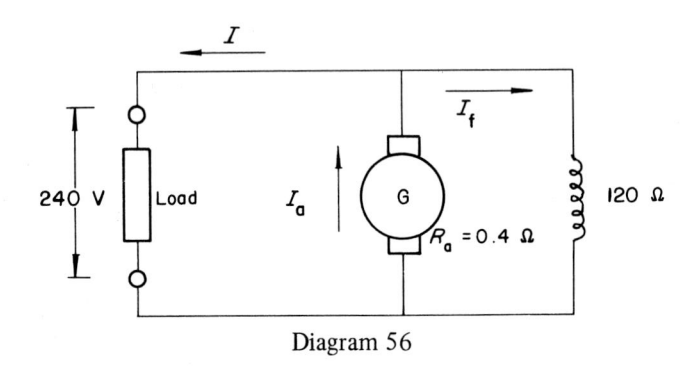

Diagram 56

As the generator supplies its own field, then from *Diagram 56*,

$$I_a = I + I_f$$

$$I_f = \frac{V}{R_f} = \frac{240}{120} = 2\text{ A}$$

$$\therefore I_a = 20 + 2 = 22\text{ A}$$

$$E = V + I_a R_a$$

$$= 240 + (22 \times 0.4)$$

$$= 240 + 8.8$$

$$= 248.8\text{ V}$$

Separately excited generator

The separately excited type of generator has its field supplied from a separate source (*Diagram 57*).

Machines 51

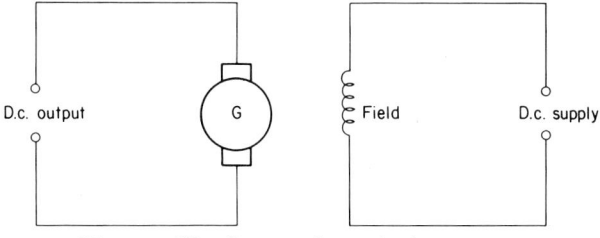

Diagram 57   Separately excited generator

Alternating-current motors

There are many different types of a.c. motor operating from either three-phase or single-phase a.c. supplies. To understand the starting problems of the single-phase types, it is best to consider three-phase motors first.

Three-phase motors

A three-phase motor depends on the rotation of a magnetic field for its movement. *Diagram 58* shows how this rotation is achieved.

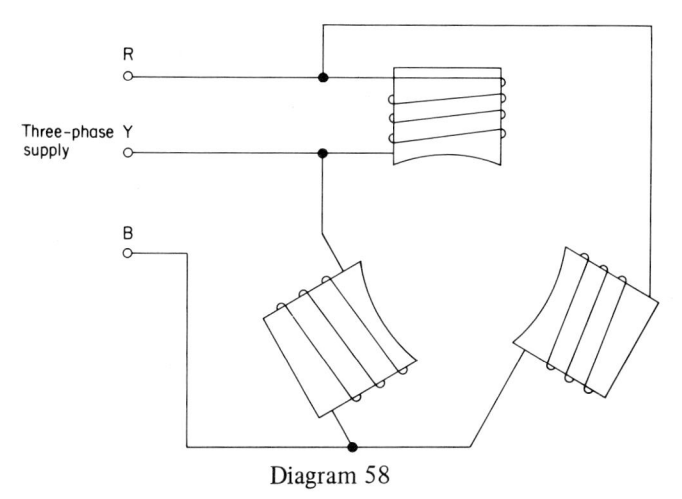

Diagram 58

If three iron-cored coils or poles are arranged 120° apart and connected as shown in *Diagram 58* to an alternating three-phase supply, then each pole will become fully energized at a different time in relation to the others. If the poles were replaced with light bulbs, it would appear as if the light were travelling around in a circular fashion from one bulb to another.

52 Machines

The iron core of each coil becomes magnetized as the coil is energized, and the arrangement gives the effect of a magnetic field rotating around the coils.

The speed of rotation of the magnetic field is called the *synchronous speed* and is dependent on the frequency of the supply and the number of *pairs* of poles, hence

$$f = np$$

where f is the supply frequency in hertz, n is the synchronous speed in revs/second and p is the number of pairs of poles.

Example 3.7

Calculate the synchronous speed of a four-pole machine if the supply frequency is 50 Hz.

$$f = np$$

$$\therefore n = \frac{f}{p}$$

$$= \frac{50}{2}$$

$$= 25 \text{ revs/second or } 1500 \text{ revs/min}$$

Synchronous motor

If we take a simple magnetic compass and place it in the centre of the arrangement shown in *Diagram 58*, the compass needle will rotate in the same direction as the magnetic field, because the magnetized compass needle is attracted to the field and therefore follows it. A three-phase synchronous motor is arranged in the same way as a three-phase generator (*Diagram 59*).

This type of motor comprises the following:

(a) a stator, which supports the magnetic field poles; and

(b) the rotor, which is basically an electromagnet supplied from a d.c. source via slip rings.

The rotor will follow the rotating magnetic field at synchronous speed.

This type of motor is not self-starting and has to be brought up to or near to synchronous speed by some means, after which it will continue to rotate of its own accord. This bringing up to speed is usually achieved by providing the rotor with some of the characteristics of an induction motor rotor.

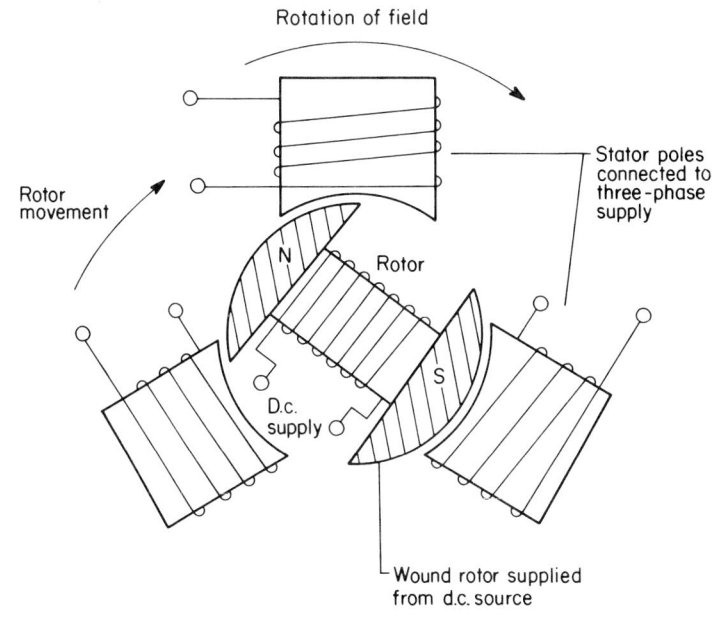

Diagram 59   Synchronous motor

Synchronous-induction motor

The synchronous-induction type of motor is essentially an induction motor with a wound rotor. It starts as an induction motor and when its speed has almost reached synchronous speed the d.c. supply is switched on and the motor will then continue to function as a synchronous motor.

This type of motor has various applications. For example, if the d.c. supply to the rotor is increased (when it is said to be 'over-excited'), the motor can be made to run at a leading power factor. This effect may be used to correct the overall power factor of an installation.

As it is a constant-speed machine, it is often used in motor-generator sets, large industrial fans and pumps.

A great advantage of the synchronous-induction type is its ability to sustain heavy mechanical overloads. Such an overload pulls the motor out of synchronism, but it continues to run as an induction motor until the overload is removed, at which time it pulls back into synchronism again.

Three-phase induction motor

Squirrel-cage type

The squirrel-cage type of induction motor comprises a wound stator and a laminated iron rotor with copper or aluminium bars embedded in it, in the form of a cage (*Diagram 60*).

54 *Machines*

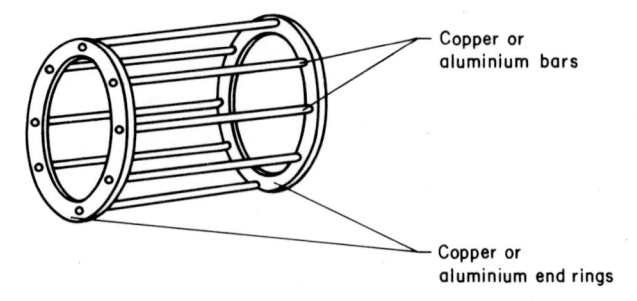

Diagram 60 Cage assembly for cage rotor

As the rotating magnetic field sweeps across the rotor, an e.m.f. is induced in the cage bars and hence a current flows. This current produces a magnetic field around the conductor and the magnetic reaction between this field and the main field causes the rotor to move. Since this movement depends on the cage bars being cut by the main field flux, it is clear that the rotor cannot rotate at synchronous speed. The speed of this type of motor is constant.

Wound-rotor type (slip ring motor)
In the wound-rotor type of motor the cage is replaced by a three-phase winding which is connected via slip rings to a starter. The starter enables the rotor currents to be controlled which in turn controls, to a degree, the speed and torque. When the machine has reached speed the rotor windings are short-circuited and the brush gear, which is no longer required, is lifted clear of the slip rings. This type of motor is capable of taking extremely high rotor currents on starting and cables must be capable of carrying such currents.

Slip
As was previously mentioned, the rotor of an induction motor cannot travel at synchronous speed, as there would be no flux cutting and the machine would not work.

The rotor is, then, said to 'slip' in speed behind the synchronous speed. Slip (S) is usually expressed as a percentage and is given by

$$\text{Slip} (\%) = \frac{(n_s - n_r)}{n_s} \times 100$$

where n_s is the synchronous speed and n_r is the rotor speed.

Example 3.8

A six-pole cage induction motor runs at 4% slip. Calculate the motor speed if the supply frequency is 50 Hz.

$$S(\%) = \frac{(n_s - n_r)}{n_s} \times 100$$

Synchronous speed $n_s = \frac{f}{p}$

$$= \frac{50}{3}$$

$$= 16.66 \text{ revs/second}$$

$$\therefore 4 = \frac{(16.66 - n_r)}{16.66} \times 100$$

$$\therefore \frac{4 \times 16.66}{100} = (16.66 - n_r)$$

$$\therefore n_r = 16.66 - \frac{(4 \times 16.66)}{100}$$

$$= 16.66 - 0.66$$

$$= 16 \text{ revs/second}$$

Example 3.9

An eight-pole induction motor runs at 12 revs/second and is supplied from a 50 Hz supply. Calculate the percentage slip.

$$n_s = \frac{f}{p}$$

$$= \frac{50}{4} = 12.5 \text{ revs/second}$$

$$S(\%) = \frac{(n_s - n_r)}{n_s} \times 100$$

$$= \frac{(12.5 - 12)}{12.5} \times 100$$

$$= \frac{0.5 \times 100}{12.5}$$

$$= \frac{50}{12.5}$$

$$= 4\%$$

Machines

Frequency of rotor currents

As the rotating field is an alternating one, the currents induced in the rotor cage bars are also alternating. These are, however, not the same frequency as the supply. The frequency of the rotor currents f_s is given by

$$f_s = \text{slip} \times \text{supply frequency}$$

$$\therefore f_s = S \times f$$

Note: S here is expressed as a per unit value; i.e., for 4% slip,

$$S = \frac{4}{100} = 0.04$$

Example 3.10

An eight-pole squirrel-cage induction motor has a synchronous speed of 12.5 revs/second and a slip of 2%. Calculate the frequency of the rotor currents.

$$f = n \times p$$
$$= 12.5 \times 4$$
$$= 50\,\text{Hz}$$

$$f_s = S \times f$$
$$= \frac{2}{100} \times 50$$
$$= 1\,\text{Hz}$$

Note: *Three*-phase motors may be reversed by changing over any *two* phases.

Single-phase induction motors

With a three-phase motor the field is displaced by 120°. In the case of a single-phase supply there is no phase displacement and hence the rotor has equal and opposing forces acting on it and there will be no movement. The motor is therefore not self-starting. However, if the rotor is initially spun mechanically it will continue to rotate in the direction in which it was turned. Of course this method of starting is out of the question with all but the very smallest motors and is therefore confined to such items as electric clocks.

The creation of an artificial phase displacement is another and more popular method of starting.

Shaded-pole induction motor

The shaded-pole type of motor has a stator with salient (projecting) poles and in each pole face is inserted a short-circuited turn of copper (*Diagram 61*).

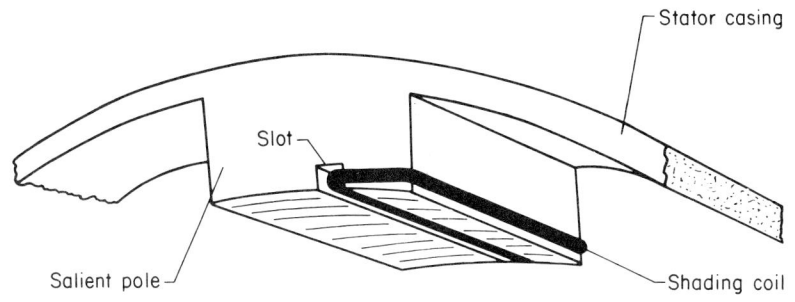

Diagram 61   Shaded-pole arrangement

The alternating flux in the pole face induces a current in the shading coil which in turn produces an opposing flux. This opposition causes a slight phase displacement of the fluxes in the two parts of each pole which is enough to start the rotor turning.

As the phase displacement is very small the motor has a very small starting torque, thus limiting its use to very light loads.

Capacitor-start induction motor

With the capacitor-start induction motor the stator has a secondary winding, in series with which is a capacitor. This gives the effect of a $90°$ phase difference and the motor will start. A second or two after starting, a centrifugal switch cuts out the secondary winding (*Diagram 62*).

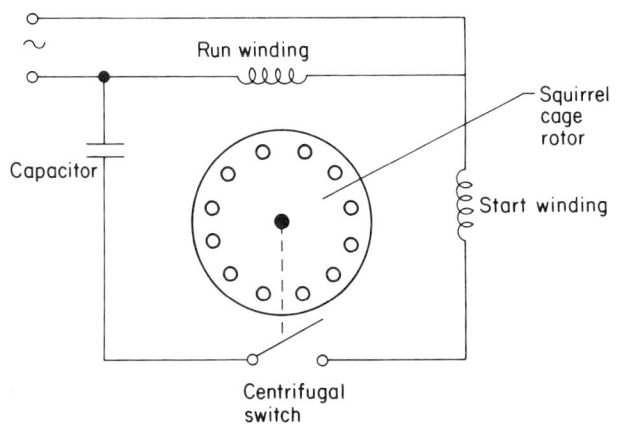

Diagram 62   Capacitor-start motor

58 *Machines*

This type of motor may be reversed in direction by reversing the connections to the start winding.

Reactance or induction-start induction motor

A phase displacement can be achieved by connecting an inductor in series with the start winding (*Diagram 63*). The centrifugal switch is as for the capacitor-start type.

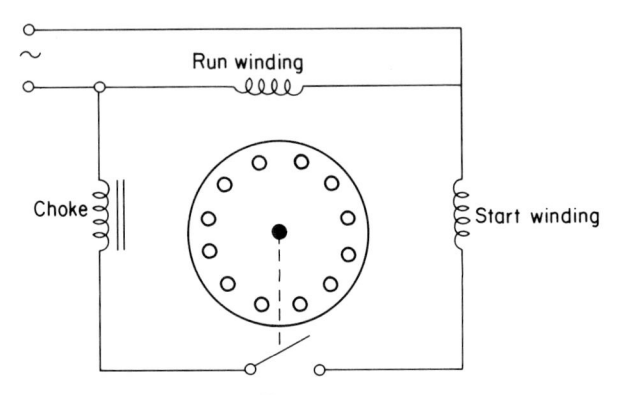

Diagram 63   Reactance-start motor

Larger motors of this type take heavy starting currents, and series resistances are used to limit this.

Reversal of rotation is as for the capacitor-start type.

Resistance-start induction motor

In the case of a resistance-start induction motor a resistance replaces the choke or capacitor in the start winding to give a phase displacement (*Diagram 64*).

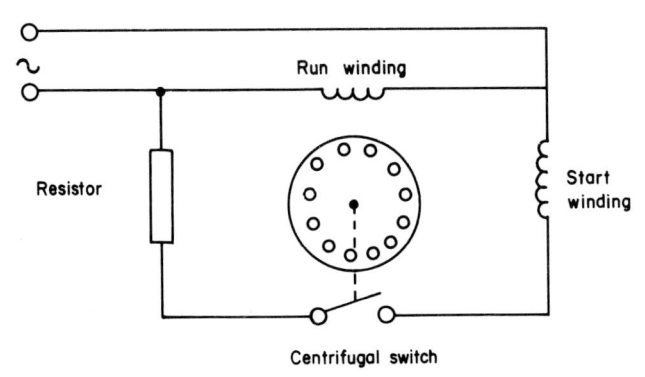

Diagram 64   Resistance-start motor

Machines 59

Capacitor-start capacitor-run induction motor

The most efficient of the range of single-phase induction motors is the capacitor-start capacitor-run type. The main feature is that the starting winding is not switched out but is continuously energized, the only change between starting and running being the value of capacitance. This change is achieved by using two capacitors and switching one out with the centrifugal switch (*Diagram 65*).

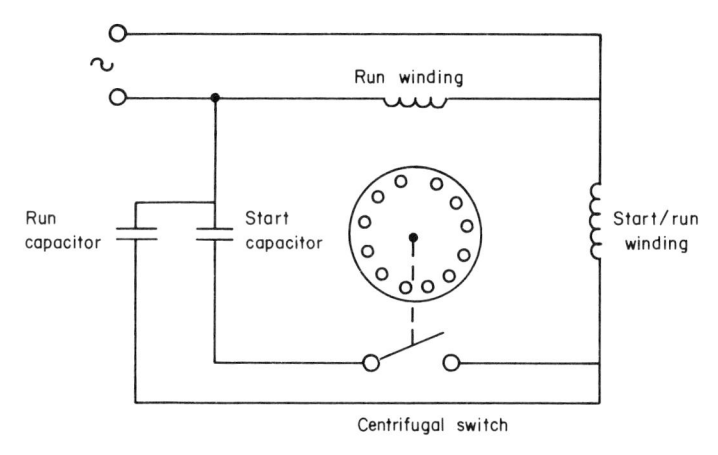

Diagram 65   Capacitor-start capacitor-run motor

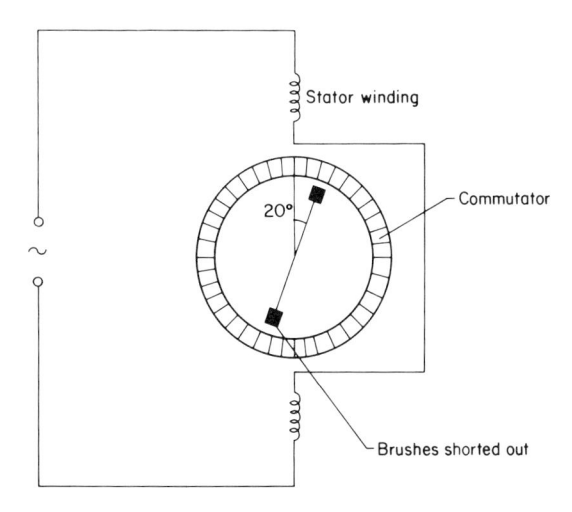

Diagram 66   Repulsion motor

Repulsion-start motor

Repulsion-start motors are of the wound-rotor type, the windings being terminated at a commutator, the brush gear of which is shorted out (*Diagram 66*) and arranged about 20° off centre. A transformer action takes place between the stator and rotor windings (mutual inductance) and as both windings will have the same polarity they repel or repulse each other. Speed control is effected by slight movement of the brushes around the commutator.

A variant of this type starts as a repulsion motor; centrifugal gear then shorts out the commutator and lifts the brush gear clear, the motor then continuing to run as an induction motor.

The universal or series motor

The *universal* or *series* motor is simply a d.c.-type armature with commutator and an a.c. field. It is connected as for a d.c. series motor (*Diagram 67*).

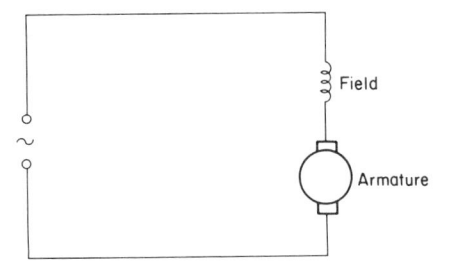

Diagram 67   Single-phase series motor

This motor will operate on an alternating current because the polarity of the a.c. supply changes on *both* field and armature; the motor will therefore rotate in one direction. Reversing is achieved by reversal of either field or armature connections.

Starters

Direct-on-line (D.O.L.) starter

Diagram 68 illustrates a typical three-phase direct-on-line starter. When the start button is pressed, the 415 V contactor coil is energized and the main and auxiliary contacts close and the motor will start. The auxiliary contact in parallel with the start button holds the coil on.

Should there be a supply voltage reduction or failure, the coil will de-energize and the motor will stop. This action provides 'no-volt' protection.

Machines 61

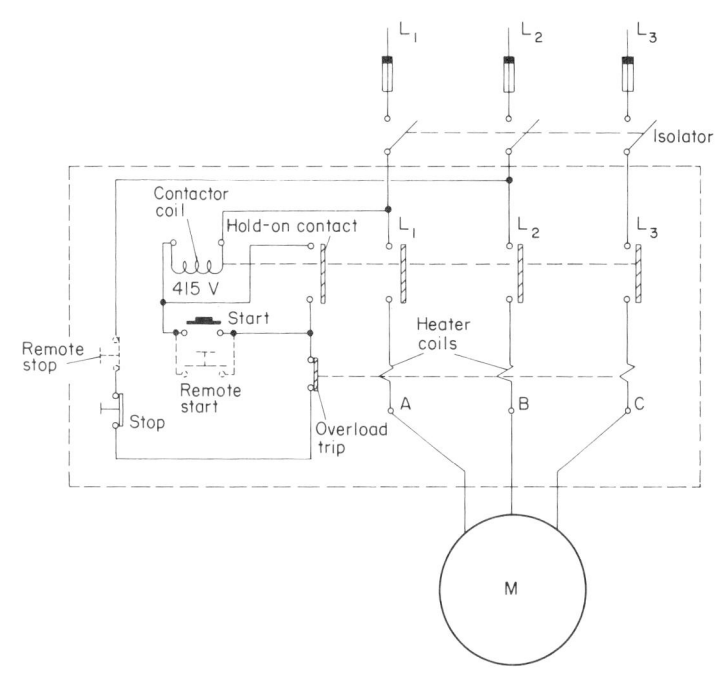

Diagram 68   Three-phase direct-on-line starter

Overload or overcurrent protection is provided by either thermal or magnetic trips.

Thermal overload protection relies on the heating effect of the load current to heat the thermal coils which in turn cause movement of a bimetallic strip. This trips out a spring-loaded contact in the control circuit. The speed at which the tripping takes place is adjusted to allow for normal starting currents, which may be four or five times as large as running currents.

Magnetic protection uses the principle of the solenoid to operate the tripping mechanism. The time lag in this case is achieved by the use of an oil or air dashpot which slows down the action of the solenoid plunger (*Diagram 69*).

Another form of thermal protection is given by the use of a *thermistor*, which is a temperature-sensitive semiconductor. It is embedded in the stator winding and activates a control circuit if the winding temperature becomes excessive.

Star-delta starter

If a motor's windings are connected in the star configuration, any two phases will be in series across the supply and hence the line current will be smaller (by 57.7%) than if the windings were connected in the delta arrangement. Hence

62 *Machines*

larger-type motors with heavy starting currents are first connected in star, and then, when the starting currents fall, in delta. This of course means that all six of the ends of the windings must be brought to terminations outside the casing (*Diagram 70*).

The automatic version of this starter incorporates a timing relay which automatically changes the connections from star to delta.

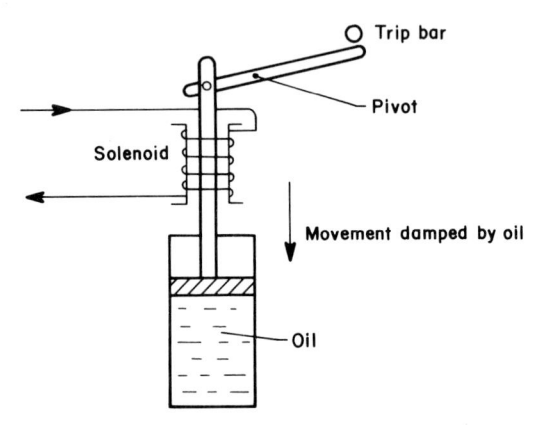

Diagram 69   Oil dashpot damping

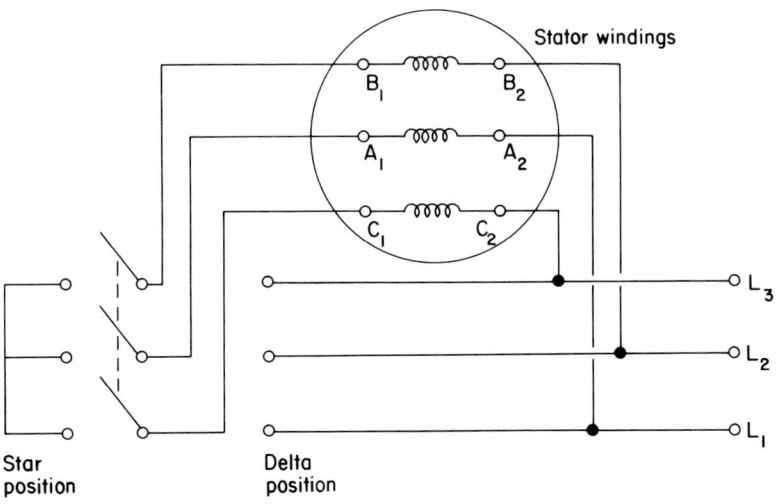

Diagram 70   Basic star–delta starter

Machines 63

Auto-transformer starting

A star-connected auto-transformer with tappings gives lower starting currents than the star–delta type (*Diagram 71*).

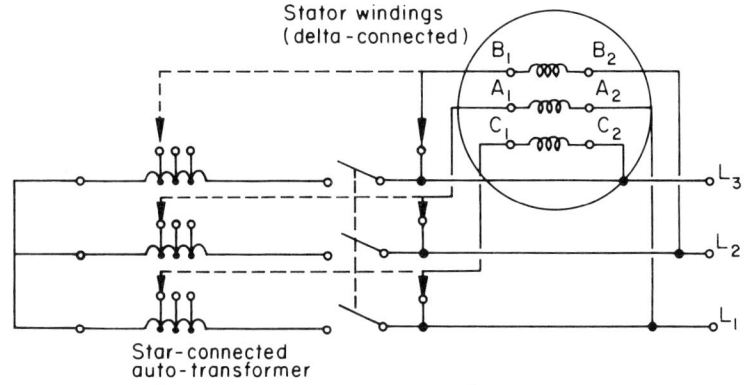

Diagram 71 Basic auto-transformer starter

Rotor-resistance starting

Diagram 72 shows the rotor resistance type starter for use with wound-rotor induction motors.

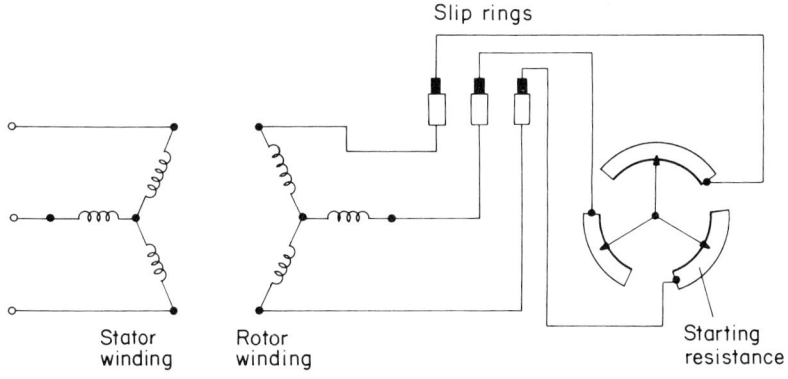

Diagram 72 Rotor resistance starter

Installing a motor

The correct handling, positioning, fixing and aligning of a motor are very important. Although it is a robust piece of machinery, great care should be taken, when transporting or positioning it, not to crack the casing or damage the feet.

Once the machine is in position it has to be fixed and aligned and this procedure will depend on the type of coupling used and the size of machine.

64   *Machines*

Very large machines are usually fixed to a concrete base or plinth. The concrete is a mixture of one part cement, two parts quartz sand and four parts of gravel. Fixing bolts should be grouted in the base with a mixture of 'one to one' washed sand and cement.

Most motors are in fact mounted on iron slide-rails, the rails being fixed to the floor, wall or ceiling depending on the motor's use. In this way, the motor can be adjusted accurately for alignment, belt tensioning and direct coupling. *Diagrams 73a, 73b* and *73c* indicate the correct methods for alignment.

Note: Always check that the insulation resistance of the motor is satisfactory before switching on the supply; dampness may have been picked up during storage.

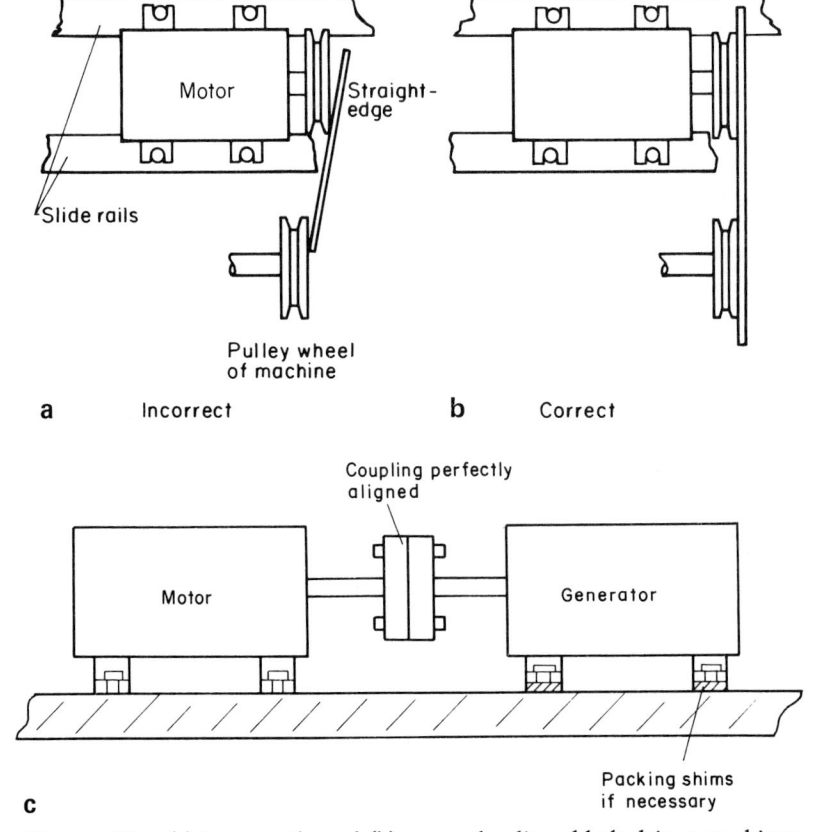

Diagram 73   (a) Incorrectly and (b) correctly aligned belt-driven machine; (c) adjustment in the case of direct coupling

Motor enclosures

Once the most appropriate type of motor for a particular task has been selected, it is necessary to ensure that the motor enclosure is also suitable for its working environment. Various kinds of enclosure are listed below and their applications are summarized in *Table 1*.

Table 1   Applications of various types of motor enclosure

Type	Applications
Screen-protected	General purposes; engineering worktops, etc.
Drip-proof	Laundries, pump rooms, etc.
Pipe-ventilated	Flour mills, cement works, paper mills, etc.
Totally enclosed	Boilerhouses, steelworks, outdoor winches, etc.
Flameproof	Gasworks, oil plants, chemical works, etc.

Screen-protected type

The most common enclosures in use are of the screen-protected type. The end covers are slotted and an internal fan draws cool air through the motor. Screen-protected enclosures can be used only in dust-free atmospheres.

Drip-proof type

The end plates on this type of motor are solid except for narrow slots on the underside. It can be used in damp and dust-free situations but is not waterproof.

Pipe-ventilated type

Cooling air is brought in via pipes from outside the building and circulated by an internal fan. This type is very suitable for extremely dusty environments.

Totally enclosed type

This type has no ventilation slots, its casing instead having ribs or fins to help cooling. The totally enclosed type is excellent for moist or dusty situations.

Totally enclosed flameproof type

This type is similar to the totally enclosed type but is more robustly built. It can withstand an internal explosion and prevent flames or sparks from reaching the outside of the casing.

Table 2 Summary of the characteristics and applications of single-phase a.c. motors

Type	Main characteristics	Applications
Universal (series)	Good starting torque. High P.F.	Small tools, drills, sanders, etc. Vacuum cleaners
Repulsion	Good starting torque. Low P.F. and efficiency. Speed control by brush shifting	Lifts, cranes, hoists, etc.
Repulsion-start	Good starting torque. Low P.F. and efficiency	Non-reversing load with heavy starting demands
Split-phase induction	Poor starting torque and P.F.	Used only with light starting conditions
Capacitor-start induction	Quite good starting torque and P.F. Quiet running	Refrigerators
Synchronous	Constant speed. Poor starting torque and P.F.	Clocks and timing devices

Table 3 Summary of the characteristics and applications of three-phase motors

Type	Main characteristics	Applications
Small synchronous, no d.c. excitation	Constant speed. Self starting. Light loads only. Low P.F.	All drives needing synchronization
Synchronous with d.c. excitation	Constant speed. Controllable P.F. High efficiency with large outputs	Compressors, P.F. correction. Ships' propulsion
Induction, squirrel-cage or wound rotor	Starting performance good for small squirrel-cage, poor for large squirrel-cage motors. Good starting with all slip-ring motors	General service in engineering, pumps, machine tools, etc.

Fault location and repairs to a.c. machines

Table 4 Possible faults on polyphase induction motors

Symptoms	Possible causes	Test and/or rectification
Fuses or over-current trips operate at start	Premature operation of protective gear Overload Reversed phase of stator winding Short-circuit or earth fault on stator circuit Short-circuit or earth fault on rotor circuit	
Motor will not start	Faulty supply or control gear Overload or low starting torque Open circuit in one stator phase Reversed phase of stator winding Open circuit in rotor circuit	
Overheated bearing or noisy operation	Bearing or mechanical defects	
Periodical growl	Reversed stator coil or coils	
Humming of squirrel-cage motor	Loose joints on rotor conductors	
Fluctuating stator current	Open circuit in rotor circuit	
General overheating of case	Faulty ventilation, mechanical or electrical overload Rotor core not fully in stator tunnel Open circuit in one of two parallel stator circuits	Reassemble motor correctly

Table 4 (continued)

Symptoms	Possible causes	Test and/or rectification
Overheating and over-labouring. Two phases of star-connected stator, or one phase of delta winding hotter than the rest	Single-phasing owing to open-circuited supply line Open circuit in one phase of stator circuit	
Reduced speed	Mechanical overload, low volts or low frequency Open circuit in rotor circuit	
Reduced speed of slip-ring motor	Rotor starter not fully operated	Overhaul protective gear to ensure correct operation
	Slip rings not short-circuited	Use slip-ring short-circuiting gear
	Volt drop on cables to rotor starter	Fit rotor starter nearer motor, or use larger rotor circuit cables

Table 5 Possible faults on single-phase induction and capacitor types of motor

Symptoms	Possible causes	Test and/or rectification
Fuses or over-current trips operate at start	Premature operation of protective gear Overload Section of stator windings reversed Short-circuit or earth fault on stator circuit Stator windings in parallel instead of series Short-circuit or earth fault on slip-ring rotor circuit	

Table 5 (continued)

Symptoms	Possible causes	Test and/or rectification
Motor will not start	Faulty supply or control circuit	
	Overload or low starting torque	
	Open circuit or reversed coils on stator winding	
	Open circuit on slip-ring rotor circuit	
	Centrifugal switch or relay sticking open	
Overheated bearing or noisy operation	Bearing or mechanical defects	
General overheating of case	Faulty ventilation, mechanical or electrical overload	
	Rotor core not fully in stator tunnel	Reassemble motor correctly
	Open circuit in one of two parallel stator circuits	
	Short-circuit on auxiliary stator winding	
	Short-circuit on centrifugal switch or relay	Overhaul switch or relay and check operation
	Centrifugal switch or relay sticking closed	Overhaul switch or relay and check operation
	Reversed section of stator windings	
	Prolonged or too frequent starting	
Reduced speed of all motors	Mechanical or electrical overload	
	Low volts or frequency	
	Open circuit in rotor	
Reduced speed of slip-ring motor	Rotor starter not fully operated	Overhaul protective gear to ensure correct operation

Table 5 (continued)

Symptoms	Possible causes	Test and/or rectification
Reduced speed of slip-ring motor (continued)	Slip-rings not short-circuited	Use slip-ring short-circuiting gear
	Volt drop on cables to rotor starter	Fit rotor starter nearer motor, or use larger rotor circuit cables

Table 6 Possible faults on series a.c. and universal motors

Symptoms	Possible causes	Test and/or rectification
Fuses or over-current trips operate at start	Premature operation of protective gear	
	Overload	
	One field coil reversed	
	Short-circuit or earth fault on field winding	
	Short-circuit or earth fault on armature	
	Wrong brush position	
Motor will not start	Faulty supply or control circuit	
	Brushes not making good contact	
	Open circuit in field windings	
	Wrong brush position	
	Short-circuit or earth fault on armature	
	Short-circuit or earth fault on field windings	
	Reversed field coil	
Overheated bearing or noisy operation	Bearing or mechanical defects	
General overheating of the case	Faulty ventilation, mechanical or electrical overload	

Table 6 (continued)

Symptoms	Possible causes	Test and/or rectification
General overheating of the case (continued)	Short-circuit, open circuit, or earth fault on armature Short-circuit or earth fault on field windings	
Reduced speed	Mechanical or electrical overload Low voltage Wrong brush position	
Increased speed	High voltage Motor unloaded	It is inadvisable to run unloaded
Sparking at brushes	Faulty brushes or commutator Mechanical or electrical overload Wrong brush position Incorrect brush spacing Open circuit, short-circuit or earth fault in armature Reversed armature coil	

Table 7 Possible faults on repulsion-type motors

Symptoms	Possible causes	Test and/or rectification
Fuses or over-current trips operate at start	Premature operation of protective gear	
	Overload	
	Section of stator windings reversed	
	Short-circuit or earth fault on stator circuit	
	Stator windings in parallel instead of series	
	Short-circuit or earth fault on armature	
	Wrong brush position	
	Commutator short-circuiting gear sticking in running position	Overhaul centrifugal gear and check operation
Motor will not start	Faulty supply or control circuit	
	Overload or low starting torque	
	Short-circuit or earth fault on stator circuit	
	Open-circuit or reversed coils on stator windings	
	Brushes not making good contact	
	Short-circuit or earth fault on armature	
	Commutator short-circuiting gear sticking in running position	Overhaul centrifugal gear and check operation
	Wrong brush position	
Overheated bearing or noisy operation	Bearing or mechanical defects	
General overheating of case	Faulty ventilation, mechanical or electrical overload	
	Rotor core not fully in stator tunnel	Reassemble motor correctly

Table 7 (continued)

Symptoms	Possible causes	Test and/or rectification
General overheating of case (continued)	Open circuit in one of two parallel stator circuits Wrong brush position Burnt contacts on commutator short-circuiting gear	Overhaul short-circuiting contacts
Overheating of repulsion–induction motor	No load	Motor usually runs hotter unloaded than on full load
Reduced speed	Low volts or frequency Wrong brush position Faulty commutator short-circuiting gear Overload Short-circuit or earth fault on armature of plain repulsion motor Open circuit in squirrel-cage of repulsion–induction motor	Overhaul short-circuiting gear. Check operation
Hunting of repulsion-start induction motor	Faulty commutator short-circuiting gear or brushes	Overhaul short-circuiting gear and brushes
Sparking at brushes	Faulty brushes or commutator Mechanical or electrical overload Wrong brush position Incorrect brush spacing Reversed armature coil Short-circuit, open circuit or earth fault on armature	

Table 8 Possible faults on synchronous types of motor

Symptoms	Possible causes	Test and/or rectification
Fuses or over-current trips operate at start	Premature operation of protective gear Overload Short-circuit or earth fault on armature	
Motor will not start	Faulty supply or control gear Low starting voltage Overload Open circuit in one armature phase	Adjust tappings on transformer
Motor fails to synchronize	External field resistance too high Open circuit in field circuit No excitation	Adjust field regulating resistor Faulty exciter
Overheated bearing or noisy operation	Bearing or mechanical defects	
Vibration	Faulty supply Open circuit in one armature phase	
General overheating	Faulty ventilation Overload High voltage Short-circuit, open circuit or earth fault on armature Incorrect field strength Unequal pole strength Unequal air gap	Adjust field regulating resistor Test field coils
Motor runs fast	High frequency	
Motor runs slow	Low frequency	
Motor pulls out of synchronism	Overload External field resistance too high	Adjust field regulating resistor

Machines 75

Table 8 (continued)

Symptoms	Possible causes	Test and/or rectification
Motor pulls out of synchronism (continued)	Open circuit in field circuit No excitation	Faulty exciter

Power factor of a.c. motors

Motors, being highly inductive pieces of equipment, have lagging power factors, some more so than others. In situations where a large number of machines are used, as in industrial premises, it is clear that some action should be taken to correct this lagging power factor. Where motors are used intermittently it is perhaps best to correct the power factor of each motor rather than the overall power factor of the installation, and capacitors connected across the terminals of each machine are used.

Example 3.11

A 240 V, 50 Hz single-phase induction motor takes a current of 13 A at a P.F. of 0.35 lagging. Calculate the value of capacitor required to correct the P.F. to 0.85 lagging.

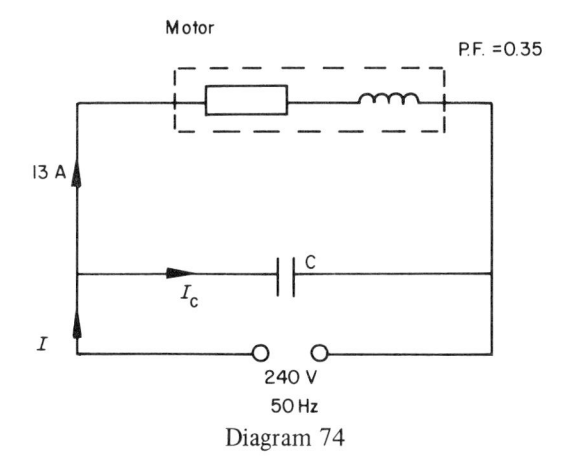

Diagram 74

The phasor diagram of the top branch (*Diagram 74*) is as shown in *Diagram 75*, while the phasor diagram of the bottom branch is as shown in *Diagram 76*. Combining both phasor diagrams so that the resultant current is at 0.85 lagging (31.8°), we have *Diagram 77*, in which

76 *Machines*

$$\cos \alpha = \text{P.F.} = 0.85$$

$$\therefore \alpha = 31.8°$$

cos θ = 0.35

240V

13 A

Diagram 75

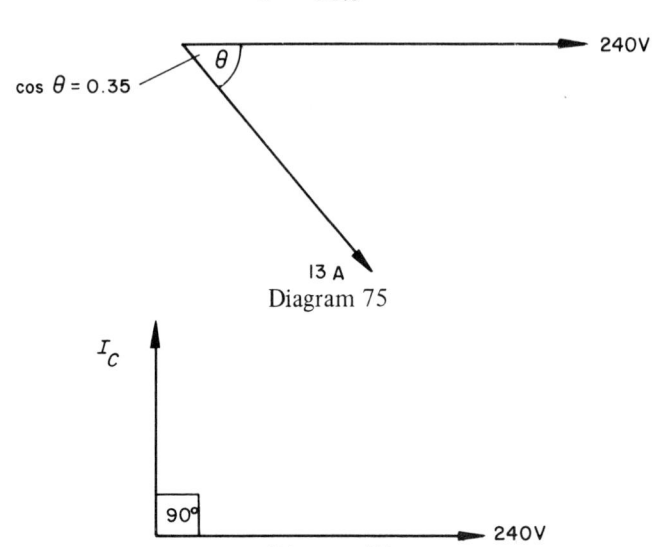

I_C

90°

240V

Diagram 76

I_C

240 V

α

I

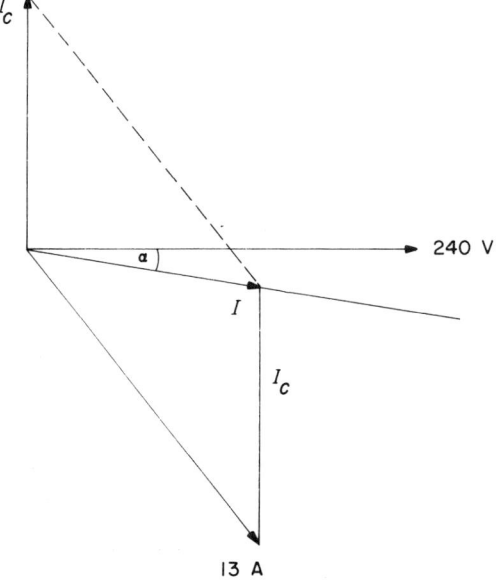

I_C

13 A

Diagram 77

By measurement,

$$I_C = 9.4 \text{ A}$$

$$X_C = \frac{V}{I_C}$$

$$= \frac{240}{9.4}$$

$$= 25.5 \, \Omega$$

$$X_C = \frac{1}{2\pi f C}$$

$$\therefore C = \frac{1}{2\pi f X_C}$$

$$= \frac{1}{314.16 \times 25.5}$$

$$= 125 \, \mu\text{F}$$

Motor ratings

As motors are a.c. plant, their electrical input is rated in kVA. The mechanical output from the machine is rated in horsepower (h.p.) or kilowatts (kW) where 1 h.p. = 746 W, and this is the rating usually displayed on the motor.

Note: Horsepower is no longer used but will still be found on older machines.

As machines have moving parts there are mechanical as well as electrical losses and they will have an efficiency given by:

$$\text{Efficiency (\%)} = \frac{\text{output} \times 100}{\text{input}}$$

Example 3.12

A 5 kW, 240 V, 50 Hz induction motor has a running P.F. of 0.7 lagging and an efficiency of 80%. Calculate the current drawn by the motor.

$$\text{Efficiency (\%)} = \frac{\text{output} \times 100}{\text{input}}$$

78 *Machines*

$$\therefore \text{Input} = \frac{\text{output} \times 100}{\text{efficiency (\%)}}$$

$$= \frac{5 \times 100}{80}$$

$$= 6.25 \text{ kW}$$

$$\text{P.F.} = \frac{\text{kW}}{\text{kVA}}$$

$$\therefore \text{kVA} = \frac{\text{kW}}{\text{P.F.}}$$

$$= \frac{6.25}{0.7}$$

$$= 8.93 \text{ kVA}$$

$$I = \frac{\text{VA}}{V}$$

$$= \frac{8.93 \times 10^3}{240}$$

$$= 37.2 \text{ A}$$

Example 3.13

A 25 kW, 415 V, 50 Hz three-phase squirrel-cage induction motor is 87% efficient and has a power factor of 0.92 lagging. Calculate the line current of the motor.

$$\text{Efficiency (\%)} = \frac{\text{output} \times 100}{\text{input}}$$

$$\therefore \text{Input} = \frac{\text{output} \times 100}{\text{efficiency (\%)}}$$

$$= \frac{25 \times 100}{87}$$

$$= 28.73 \text{ kW}$$

$$\text{Power (watts)} = \sqrt{3} \ V_L \times I_L \times \text{P.F.}$$

$$\therefore I_L = \frac{P}{\sqrt{3}\ V_L \times \text{P.F.}}$$

$$= \frac{28\ 730}{\sqrt{3} \times 415 \times 0.92}$$

$$= 43.45\ \text{A}$$

Torque and output

Example 3.14

A four-pole cage induction motor is run from a 50 Hz supply and has a slip of 3%. The rotor shaft drives a pulley wheel 300 mm in diameter, which has a tangential force of 200 N exerted upon it. Calculate the power output from the rotor in watts.

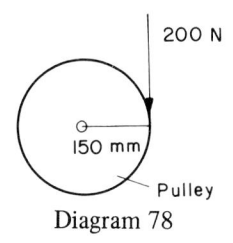

Diagram 78

From *Diagram 78*,

$$\text{Torque} = \text{force} \times \text{radius}$$

$$= 200 \times 150 \times 10^{-3}$$

$$= 30\ \text{Nm}$$

$$\text{Slip}(\%) = \frac{(n_s - n_r)\,100}{n_s}$$

and

$$n_s = \frac{f}{p}$$

$$= \frac{50}{2}$$

$$= 25\ \text{revs/second}$$

80 Machines

$$\therefore 3 = \frac{(25 - n_r)100}{25}$$

$$\therefore \frac{3 \times 25}{100} = 25 - n_r$$

$$\therefore n_r = 25 - \frac{(3 \times 25)}{100}$$

$$= 25 - 0.75$$

$$= 24.25 \text{ revs/second}$$

$$P = 2\pi n T$$

$$= 2\pi \times 24.25 \times 30$$

$$= 4.57 \text{ kW}$$

Points to note (I.E.E. Regulations)

1. The voltage drop between the supply intake position and the motor must not exceed 2.5% of the supply voltage.
2. The motor enclosure must be suitable for the environment in which it is to work; for example, a flameproof enclosure is required for explosive situations.
3. Every motor must have a means of being started and stopped, the means of stopping being situated within easy reach of the person operating the motor.
4. Every motor should have a control such that the motor cannot restart after it has stopped because of mains volt drop or failure (i.e. under-voltage protection). This regulation may be relaxed if a dangerous situation will arise should the motor fail to restart.
5. *Every* stopping device must have to be reset before a motor can be restarted.
6. A means of isolation must be provided for every motor and its associated control gear. If this means is remote from the motor, isolation adjacent to the motor must be provided, or the remote isolator must be capable of being 'locked off'.
7. If a single motor and/or its control gear in a group of motors is to be maintained or inspected, a single means of isolation for the whole group may be installed provided the loss of supply to the whole group is acceptable.

8. Excess current protection must be provided in control gear serving motors rated above 370 watts and/or in the cables between the protection and the motor.
9. Cables carrying the starting and load currents of motors must be at least equal in rating to the full-load current rating of the motor. This includes rotor circuits of slip-ring or commutator motors.
10. The final circuit supplying a motor shall be protected by fuses or circuit breakers of rating not greater than that of the cable, unless a starter is provided that protects the cable between itself and the motor, in which case the fuses or circuit breakers may be rated up to twice the rating of the cable between the fuse and the starter (*Diagram 79*).

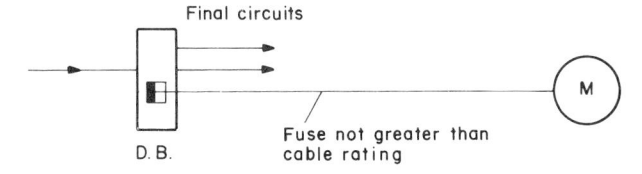

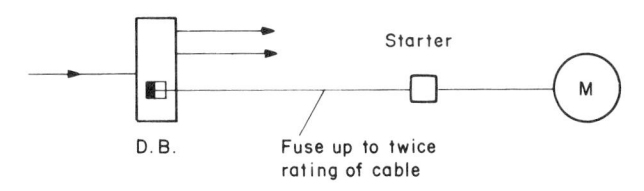

Diagram 79   (D.B. = distribution board)

Questions on Chapter 3

1. (a) Explain what is meant by the term 'back e.m.f.' in a motor.
   (b) Outline the basic differences between series, shunt and compound-wound d.c. motors.
2. (a) With the aid of diagrams explain how the speed of series, shunt and compound-wound motors may be controlled.
   (b) A 440 V d.c. shunt-wound motor has a field resistance of 200 Ω and an armature resistance of 0.6 Ω. When its speed is 20 revs/second the current drawn from the supply is 12.2 A. Calculate its back e.m.f. at

this speed. If the speed were decreased to 19 revs/second, the field flux remaining unchanged, calculate the new back e.m.f. and the new armature current.

3. A load of 6.4 kW at 240 V is supplied from the terminals of a shunt-wound d.c. generator. The field resistance is 180 Ω. Calculate the armature current.

4. (a) Explain with the aid of a diagram how a rotating magnetic field may be obtained.
   (b) What is meant by 'synchronous speed'? Calculate the synchronous speed of a 12 pole motor if the supply frequency is 50 Hz.

5. (a) Explain the action of a synchronous motor.
   (b) What methods are available to start a synchronous motor? Explain with diagrams.

6. (a) Explain the action of a cage induction motor.
   (b) What is meant by the term 'slip'? Calculate the percentage slip of a six-pole induction motor running at 16.2 revs/second from a 50 Hz supply.

7. Describe with sketches three different ways of starting a single-phase induction motor.

8. Explain with the aid of a sketch the action of a three-phase D.O.L. starter. What are dashpots used for? How are remote start and stop buttons connected?

9. A 10 kW, 240 V, 50 Hz single-phase cage rotor induction motor is 85% efficient and has a P.F. of 0.68 lagging. Calculate the motor current and the value of capacitor required to raise the P.F. to 0.93 lagging.

10. Calculate the torque developed by an 18 kW four-pole induction motor run at 3.5% slip from a 50 Hz supply.

CHAPTER 4

Rectification

Semiconductors

All matter is made up of atoms. Materials having atoms with many 'easily dislodged' electrons are called *conductors*, and those with very few such electrons are called *insulators*. Somewhere in between these two extremes lie materials which are neither conductors nor insulators. They are known as *semiconductors*.

These semiconductor materials have only a few electrons in their outer shells or valence bands and are easily made to accept or give up electrons. Such materials include selenium, copper oxide, germanium and silicon. The first two are used in metal rectifiers, the second two in junction rectifiers, silicon being the more widely used.

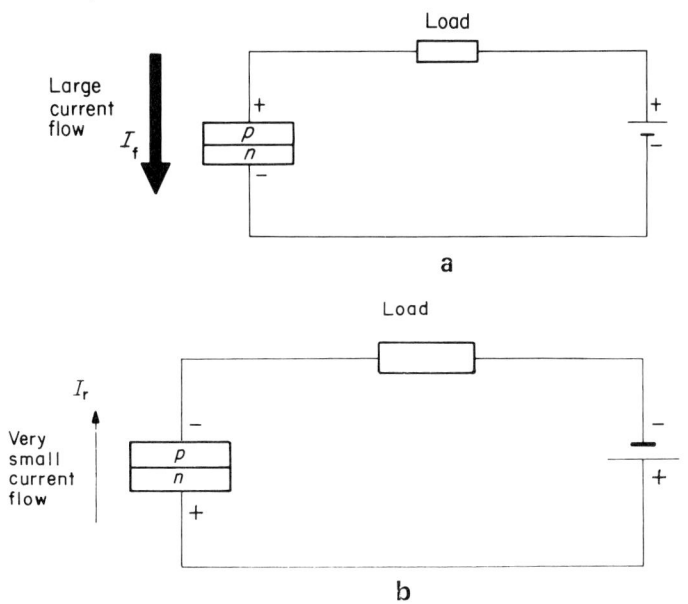

Diagram 80 (a) Diode conducts, forward bias; (b) diode insulates, reverse bias

84 *Rectification*

If tiny amounts of impurities such as arsenic or aluminium are added to samples of silicon, the samples can be made to lose electrons and become positively charged (*p*-type silicon) or gain electrons and become negatively charged (*n*-type).

By placing two different samples together (a *diode*) and applying a voltage across them, the assembly will act as a conductor. Reversal of the polarity (reverse voltage) will make the samples act as an insulator (*Diagrams 80a* and *80b*).

The symbol for the diode is shown in *Diagram 81*. The diode will conduct when current flow is in the direction of the arrow.

Diagram 81

Diagram 82 shows how the forward and reverse current for a silicon diode vary with the applied voltage. At point X in the graph the reverse voltage is so great that the diode breaks down and conducts. This value is called the reverse breakdown voltage.

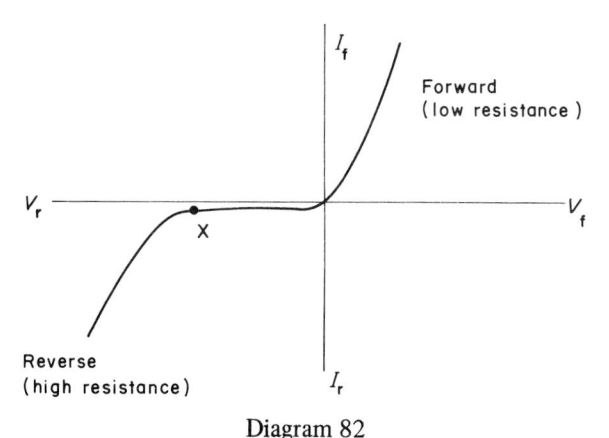

Diagram 82

Rectification

Most electricity supply systems are a.c. and since many appliances require a d.c. supply, it is necessary to change a.c. to d.c. This change is called *rectification*.

Rectification 85

Diagrams 83a, 83b and *83c* illustrate how a diode or group of diodes can be used to rectify an a.c. supply. It can be seen that the rectified d.c. output is not true d.c., for which the waveform would be a straight line, but has something of a pulsating nature. This type of output is usually quite acceptable for most purposes in installation work. Should a more refined or smoothed output be required, the addition of capacitance and inductance (*Diagram 84*) can provide this.

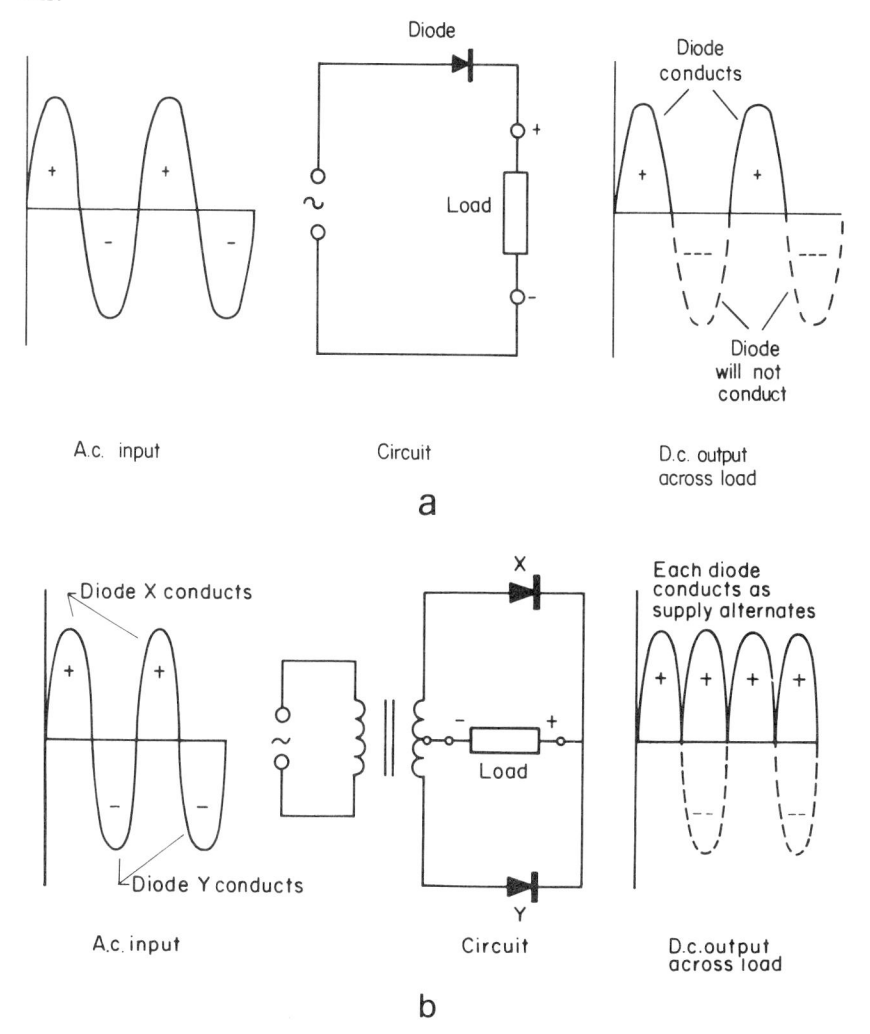

Diagram 83 (a) Half-wave rectification; (b) full-wave rectification from transformer supply

86 Rectification

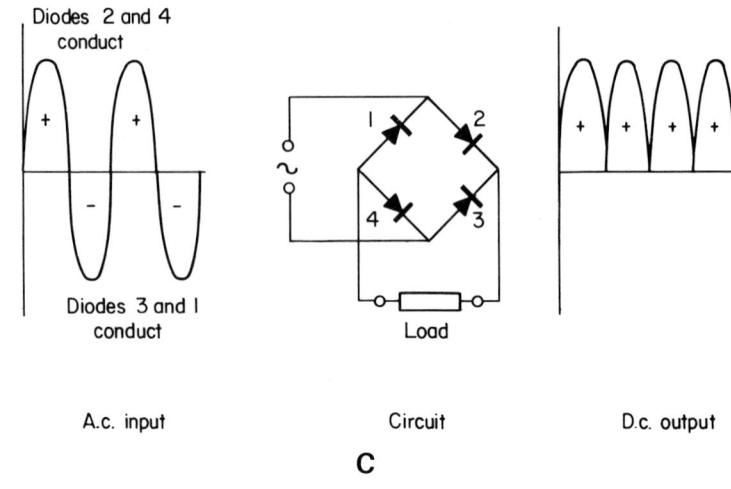

A.c. input — Diodes 2 and 4 conduct / Diodes 3 and 1 conduct

Circuit

D.c. output

Diagram 83 (c) Bridge-type full-wave rectification

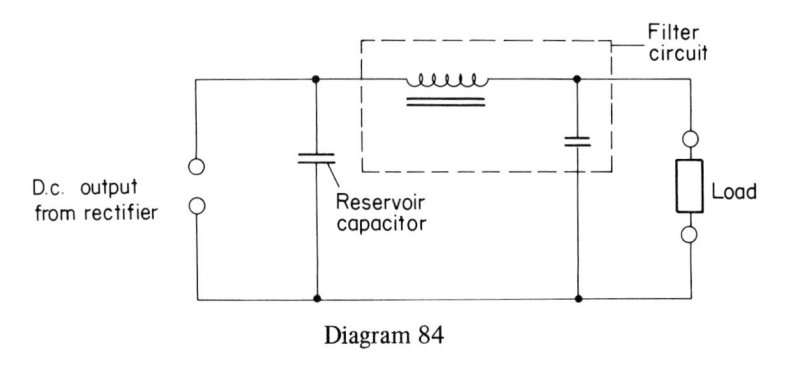

Diagram 84

Rectifier output

Since the output from a full-wave rectifier is a series of sinusoidal pulses, the average value of this output is given by

$$\text{Average value} = \text{maximum or peak value} \times 0.637$$

and for half-wave

$$\text{Average value} = \frac{\text{maximum or peak value} \times 0.637}{2}$$

Note: When selecting a diode for a particular duty, ensure that it is capable of operating at the peak voltage. For example, consider a circuit that is required to operate at 12 V d.c. and a 240/12 V transformer is to be used in conjunction with diodes.

The 12 V output from the transformer is 12 V **r.m.s.** and therefore has a peak value of

$$\frac{12}{0.7071} = 16.97 \text{ V}$$

The diode must be able to cope with this peak voltage.

Example 4.1

Calculate the average value of a full-wave rectified d.c. output if the a.c. input is 16 V.

$$\text{Peak value of a.c. input} = \frac{16}{0.7071} = 22.61 \text{ V}$$

$$\text{Average value of d.c. output} = 22.61 \times 0.637 = 14.4 \text{ V}$$

Applications

1. D.c. machines (supply).
2. Bell and call systems.
3. Battery charging.
4. Emergency lighting circuits.

Heat sinks

Semiconductor devices become extremely hot whilst in use and unless controlled, this heat can cause damage.

Usually the devices are located in the centre of a sheet of metal, the heat being transferred from the semiconductor to the sheet and so to the surrounding air. The process is similar to the action of cooling fins for transformers.

Thyristors or silicon-controlled rectifiers

A thyristor is a four-layer *p-n-p-n* device with three connections (*Diagram 85*).

88   *Rectification*

Diagram 85   Thyristor

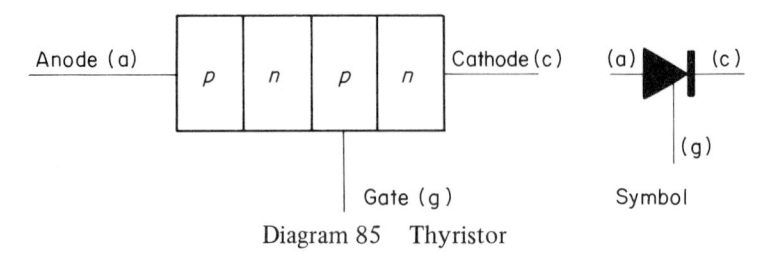

Under normal circumstances, with positive on the anode and negative on the cathode, the thyristor will not conduct. If, however, a large enough positive firing potential is applied to the 'gate' connection, the thyristor will conduct and will continue to do so even if the signal on the gate is removed. It will cease to conduct when the anode potential falls below that of the cathode. The device, like the diode, will not conduct at all in the reverse direction.

Let us now consider what happens when a thyristor is wired in an a.c. circuit as shown in *Diagram 86*. Resistor R_1 ensures that the minimum gate potential required for firing is maintained and R_3 ensures that the gate potential does not rise to a level that could cause damage to the gate circuit. Variable resistor R_2 enables various gate potentials to be selected between maximum and minimum, and it will also be seen that as the circuit is resistive, the applied voltage and the gate voltage are in phase (*Diagram 87*).

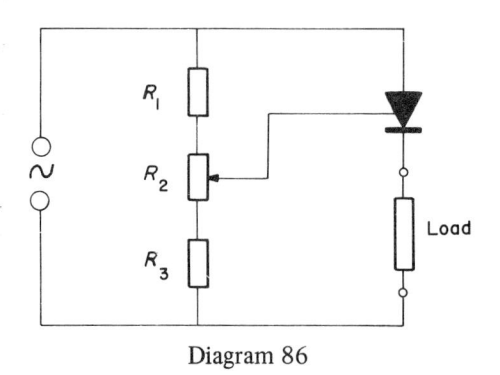

Diagram 86

It will be seen from *Diagram 87* that if the gate voltage is adjusted to a maximum, the gate will fire at point A. When it is at a minimum it fires at point B.

The current flowing in a resistive circuit is also in phase with the voltage and in this case it will flow only when the thyristor is conducting. Since this conduction takes place only when the thyristor is triggered by the gate, the amount of current flowing in any +ve half cycle can be controlled by the gate potential (*Diagrams 88a* and *88b*).

Diagram 87

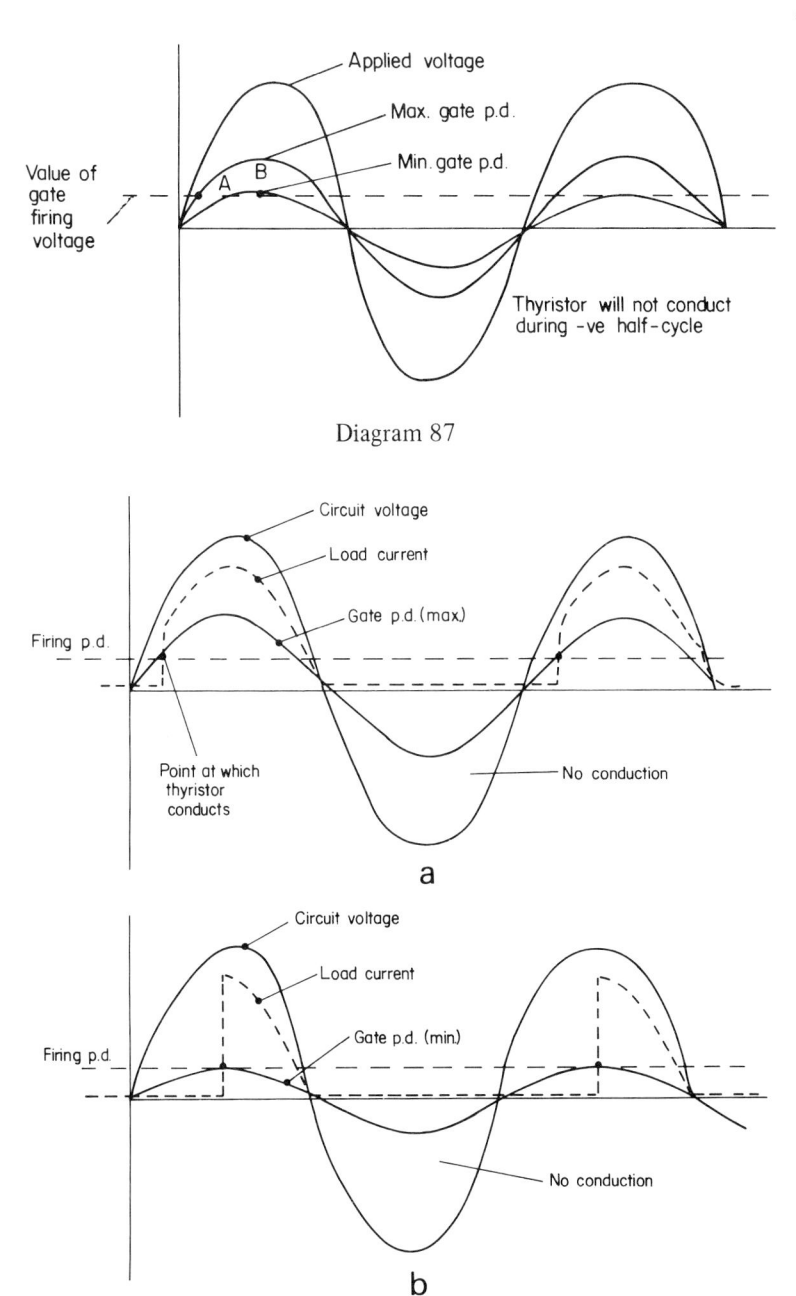

Diagram 88 (a) Maximum gate potential; (b) minimum gate potential

90 *Rectification*

This control of the amount of current flowing each half cycle can be used to control the speed of small motors such as those used in food mixers and hand drills. A simple circuit is shown in *Diagram 89*.

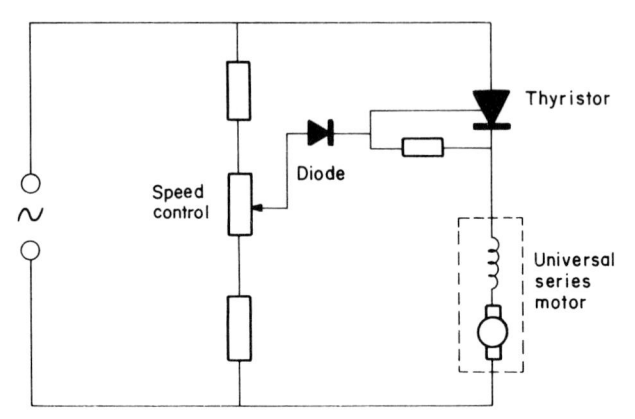

Diagram 89   Basic speed control circuit for a motor

More complicated circuitry is now in use, utilizing thyristors, in order to control the speed of induction motors, something that in the past proved very difficult.

CHAPTER 5
Earthing

If metalwork that is not earthed comes into contact with a live supply, there is a serious risk that someone may suffer an electric shock. The metalwork rises to a dangerously high potential and a person coming into contact with it and earth could, depending on circumstances, receive a lethal shock.

In order to prevent this situation from arising, the I.E.E. Regulations recommend the following alternatives:

(a) ensure that the apparatus of which the metalwork is part, is of 'all insulated' construction; or

(b) it is of the double-insulated type; or

(c) ensure that the metalwork is isolated so that it cannot come into contact with live parts or earthed metal; or

(d) earth all exposed metal parts.

It is the last of these alternatives that will be considered here.

The Regulations do not confine earthing to metalwork of apparatus containing live conductors. Included is all metalwork that could come into accidental contact with live conductors, and this includes baths, sinks, metal radiators, and similar items.

Exposed metalwork forming part of an electrical installation is known as an 'exposed conductive part', whereas metalwork such as gas, water and oil pipes and structural steelwork is known as an 'extraneous conductive part'.

Earth-fault loop path

In order for protective devices to operate under earth fault conditions, it is necessary for a complete circuit to exist for the passage of fault current. This circuit is called the earth-fault loop path and is shown in *Diagram 90*.

In many installations the earthing conductor is connected to the supply cable metal sheath, and the earth return is vai that metallic path (T.N.-S. system). However, in a great many rural areas where supplies are by overhead cable, no metallic return path is available and the general mass of earth must be relied on, connection to earth being via electrodes driven into the ground (T.T. system).

92 *Earthing*

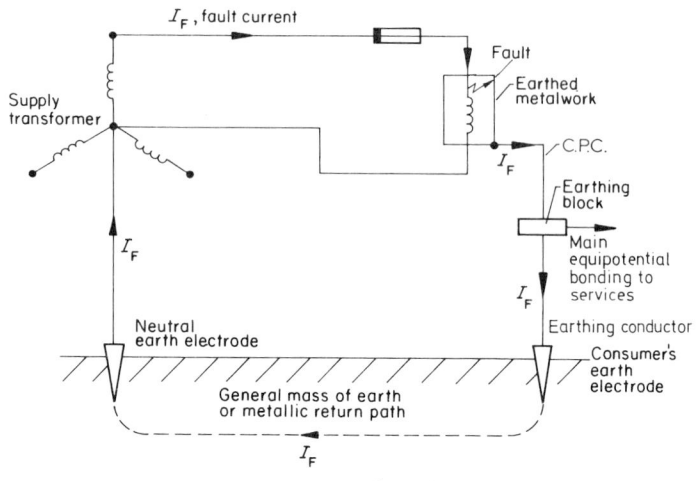

Diagram 90   Earth-fault loop path

Loop impedance

The value of the opposition to the flow of fault current in the loop path is of great importance. The higher the opposition, the smaller the current available to operate the protection. Since the path is made up of the resistance of the phase conductor, the earthing conductors, the mass of earth and the resistance and reactance of the transformer winding, the opposition is termed 'impedance' rather than 'resistance', and the Regulations quote maximum values of impedance for various types of protection for socket outlet circuits and for circuits feeding fixed equipment. For example a radial socket outlet circuit protected by a 20 A rewirable fuse to B.S. 3036 should have an associated loop impedance Z_S of not more than 1.8 Ω. This means that the current that would flow under fault conditions would be:

$$I_F = \frac{U_0}{Z_S} = \frac{240}{1.8} = 133.\dot{3} \text{ A (where } U_0 = \text{voltage to earth)}$$

which should operate the fuse in the required 0.4 s. If it were not possible to obtain the required value of Z_S, it may be necessary to replace the fuse by a type 2 circuit breaker to B.S. 3871, i.e. one which has a thermal trip for overloads and a magnetic trip for short circuits. In this case the value of Z_S would be increased to 3 Ω

hence $$I_F = \frac{240}{3} = 80 \text{ A}$$

Since circuits feeding fixed equipment are less liable to shock risk, the operating or disconnection time of the protection can be increased to 5 s, which means that the operating current of the protection can be smaller, and hence the loop impedance value will be higher. For example for a 20 A fuse to B.S. 3036 protecting fixed equipment the value of Z_S is 4 Ω, giving:

$$I_F = \frac{240}{4} = 60 \text{ A}$$

However, loop impedances lower than 10 Ω using the earth as a return path, are hard to achieve, due to the usually high value of electrode to earth resistance.

Earth-electrode resistance

If we were to place an electrode in the earth and then measure the resistance between the electrode and points at increasingly larger distances from it, we would notice that the resistance increased with distance until a point was reached (usually around 2.5 m) beyond which no increase in resistance was seen (*Diagram 91*).

The value of this *electrode resistance* will depend on the length and cross-sectional area of the electrode and the type of soil.

This resistance area is particularly important with regard to voltage at the surface of the ground (*Diagram 92*).

For a 2 m earth rod, with its top at ground level, 80% to 90% of the voltage appearing at the electrode under fault conditions is dropped across the earth in the first 2.5 to 3 m. This is particularly dangerous where livestock are present as the hind and fore legs of an animal can be respectively inside and outside the resistance area: 25 V can be lethal. This problem can be overcome by ensuring that the whole of the electrode is well below ground level and by providing protection that will operate in a fraction of a second (earth-leakage circuit breaker).

As the earth-electrode resistance is by far the largest part of the loop impedance, it is clearly necessary, when installing such an electrode, to measure the earth resistance.

Measurement of earth-electrode resistance

To conduct the test it is necessary to disconnect the electrode under test from the earthing system. The basic concept of the test is shown in *Diagram 93*. An electrode C is driven in the earth at a distance of some 30 to 50 m from electrode A, and electrode B is placed approximately half-way between them. This ensures that B is outside the resistance area of A.

An alternating current from a hand-held generator is applied to the electrodes as shown in *Diagram 93*, the resistance being calculated from the

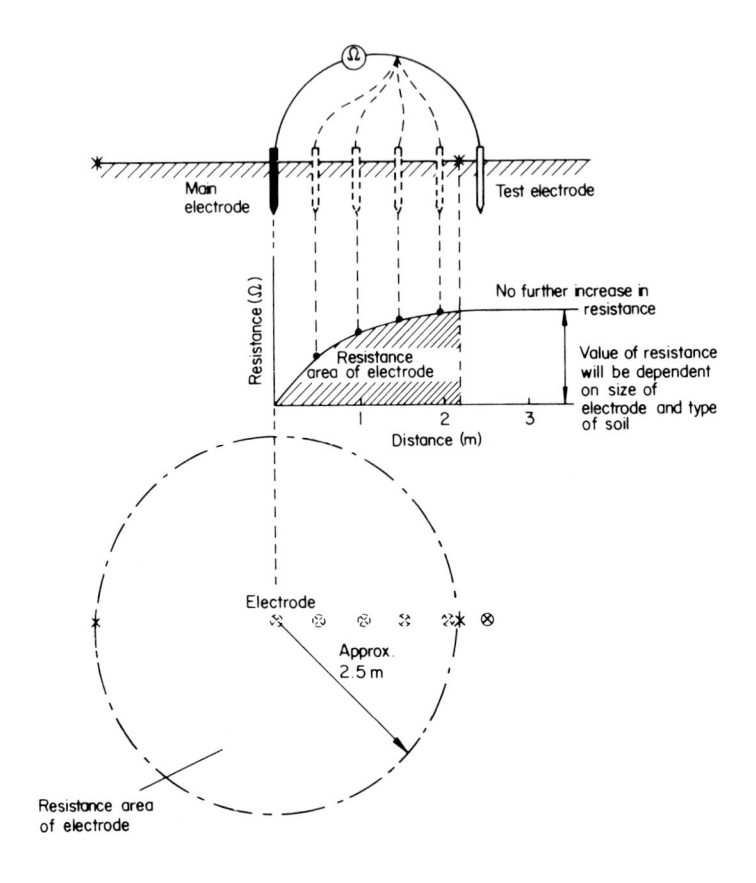

Diagram 91   Resistance area of electrode

Earthing 95

```
         240 V              25 V
```

Diagram 92

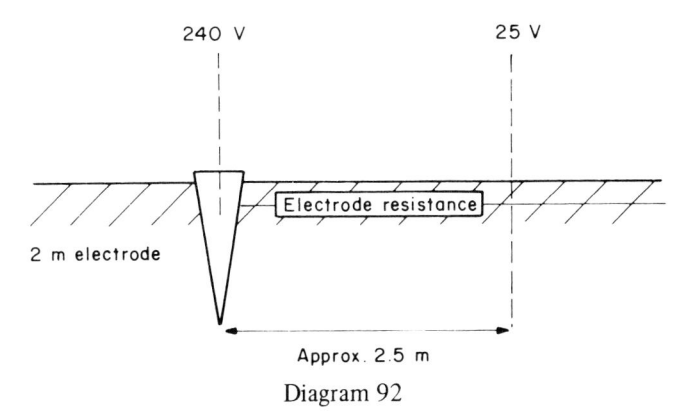

values of current and voltage (most commercial instruments provide a direct-reading ohmmeter).

Electrode B is then moved first to position B_1 then to B_2 6 m either side of position B and the readings taken again. Provided the three readings are similar, this value may be taken as the earth-electrode resistance.

Of course, the values of resistance recorded will depend on the type of soil and its moisture content. Any soil that holds moisture such as clay or marshy ground has a relatively low resistivity, whereas gravel or rock have a high resistivity. Typical values for soil resistivity in ohms are as follows:

Garden soil	5 to 50
Clay	10 to 100
Sand	250 to 500
Rock	1 000 to 10 000

Calculation of fault current and voltage

This is best illustrated by an example.

Example 5.1

An installation is earthed via an earth electrode of resistance 12 Ω and a supply authority neutral electrode of resistance 18 Ω. A fault develops on the 30 A cooker circuit, the insulation being damaged such that the live-earth insulation resistance is only 10 Ω. Draw a diagram of the fault circuit and calculate the fault current, and the voltage on the consumer's earthing system. (Ignore the reactance of the transformer.)

96 *Earthing*

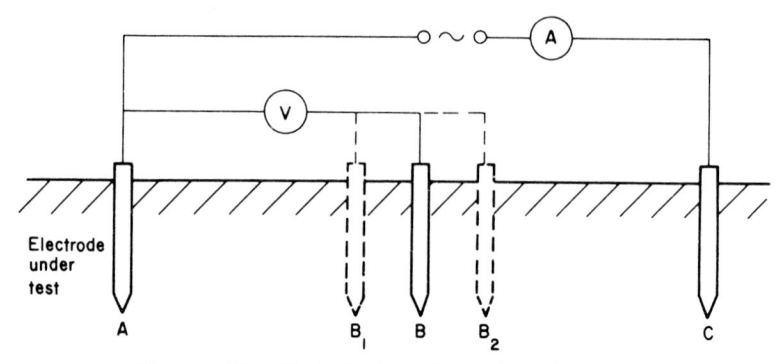

Diagram 93 Test of earth-electrode resistance

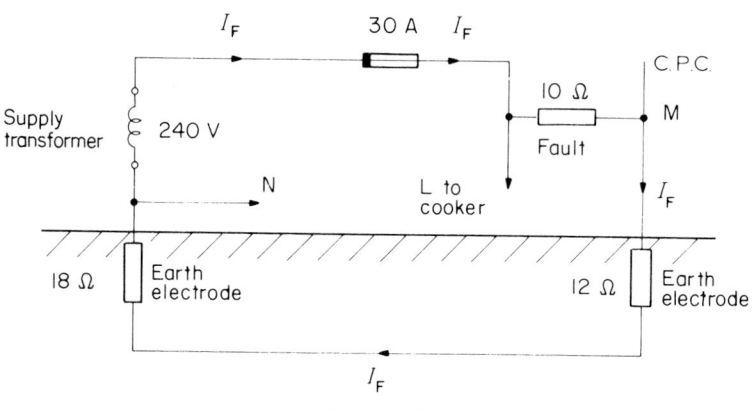

Diagram 94

From *Diagram 94*,

Total resistance of fault path = 10 + 12 + 18 = 40 Ω

(Resistance of phase conductor and C.P.C. is ignored.)

Earthing 97

$$\therefore \text{Fault current } I_F = \frac{240}{40} = 6\,\text{A}$$

Potential on circuit protective conductor (C.P.C.) = 240 − volt drop across fault

$$\text{Fault volt drop} = I_F \times 10$$
$$= 6 \times 10$$
$$= 60\,\text{V}$$
$$\therefore \text{Potential on C.P.C.} = 240 - 60$$
$$= 180\,\text{V}$$

This example shows that with these typical values of earth-electrode resistance a current of 6 A will flow uninterrupted and a potentially lethal voltage of 180 V will remain on the consumer's earthing system.

In order to ensure protection against danger from earth leakage currents in areas depending on the general mass of earth, residual current circuit breakers (r.c.c.b.s) are used. As these devices do not rely on a connection in the earth circuit, they have replaced the fault voltage type, which are no longer permitted.

Protective multiple earthing (TN-C-S system)

Another method of providing an earth return path is to make use of the neutral conductor. Here the installation earth is connected to the neutral conductor at the service position. This converts live-earth faults to live-neutral faults and allows heavy currents to flow which will operate the protection. With this system ('protective multiple earthing', P.M.E.) it is important to ensure that the neutral is kept at earth potential by earthing it at many points along its length (hence 'multiple' earthing) (*Diagram 95*). If this is not done, a fault to neutral in one installation could cause a shock risk in all the other installations connected to that system.

The cable usually used for such a system is concentric cable which consists of a single-core cable (for single-phase) surrounded by armouring which is the earth and neutral conductor (*Diagram 96*). Three-core with concentric neutral would be used for three-phase, four-wire cable.

There are, however, several hazards associated with the use of a P.M.E. system. These include:

98 *Earthing*

1. Shock risk if neutral is broken. In this case, a fault on a P.M.E. system with a broken neutral would result in the neutral becoming live to earth either side of the break. This situation is more likely to occur with overhead supplies.
2. Fire risk. As heavy currents are encouraged to flow, there is a risk of fire starting during the time it takes for the protective devices to operate.

The chance of a broken neutral is lessened in underground cable to some extent by the use of concentric cable, as it is unlikely that the neutral conductor in such a cable could be broken without breaking the live conductor.

In view of the hazards of such a system there are strict regulations for its use, and approval from the Department of Energy must be obtained before it can be installed.

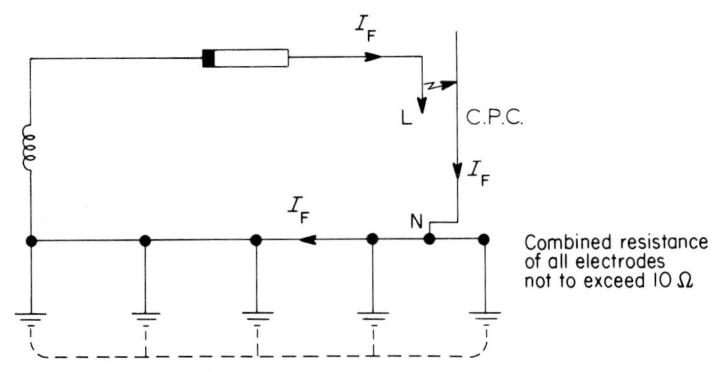

Diagram 95   Protective multiple earthing or TN-C-S system

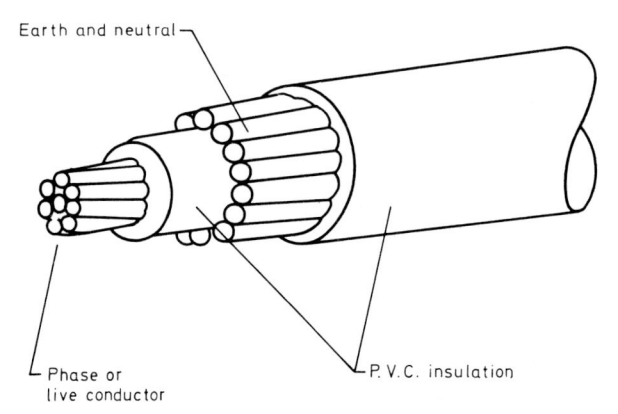

Diagram 96   Single-core concentric cable

Of course, providing a good earth return path does not ensure that a consumer's premises are protected against dangerous leakage currents unless his earthing system is intact and connected to the supply authority's earth. This possible fault situation can be overcome by using an earthed concentric wiring system in the premises.

Earthed concentric wiring

Earthed concentric wiring, or combined neutral and earth (C.N.E.), is simply an extension of the P.M.E. system into the consumer's premises. The installation cable is mineral-insulated metal-sheathed, the outer sheath containing earth and neutral, which is called the P.E.N. conductor. In this way the neutral and hence C.P.C. must be continuous and connected to the supply neutral, otherwise the whole or parts of the installation would not work.

To ensure continuity of the sheath throughout an installation, *two* separate earth bonds must be fixed across each break in the sheathing, at socket outlets and/or lighting points. This is usually achieved with the use of a sealing pot with a special earth tail fixed to it and the connection via the metal box and cable glands (*Diagrams 97a* and *97b*).

Diagram 98 shows a typical circuit for a light operated from a one-way switch.

Although the earthed concentric wiring system provides a satisfactory consumer's earthing arrangement, it is an expensive system to install and still suffers from the hazards of broken neutrals and heavy fault currents. Also, in common with other systems using solid earthing, it cannot detect and protect against high resistance faults. This may not cause a fire or shock risk but will cause a higher electricity bill.

Example 5.2

A consumer has an undetected live-earth fault on a cooker circuit. The fault resistance is 100 Ω. Calculate the cost to the consumer over a one-year period at 2.7p per unit.

$$\text{Current drawn from supply} = \frac{240}{100} = 2.4 \text{ A}$$

Clearly the cooker circuit protection will not operate.

$$\begin{aligned}
\text{Power consumed} &= I_F^2 \times R_F \\
&= 2.4^2 \times 100 \\
&= 5.76 \times 100 \\
&= 576 \text{ W}
\end{aligned}$$

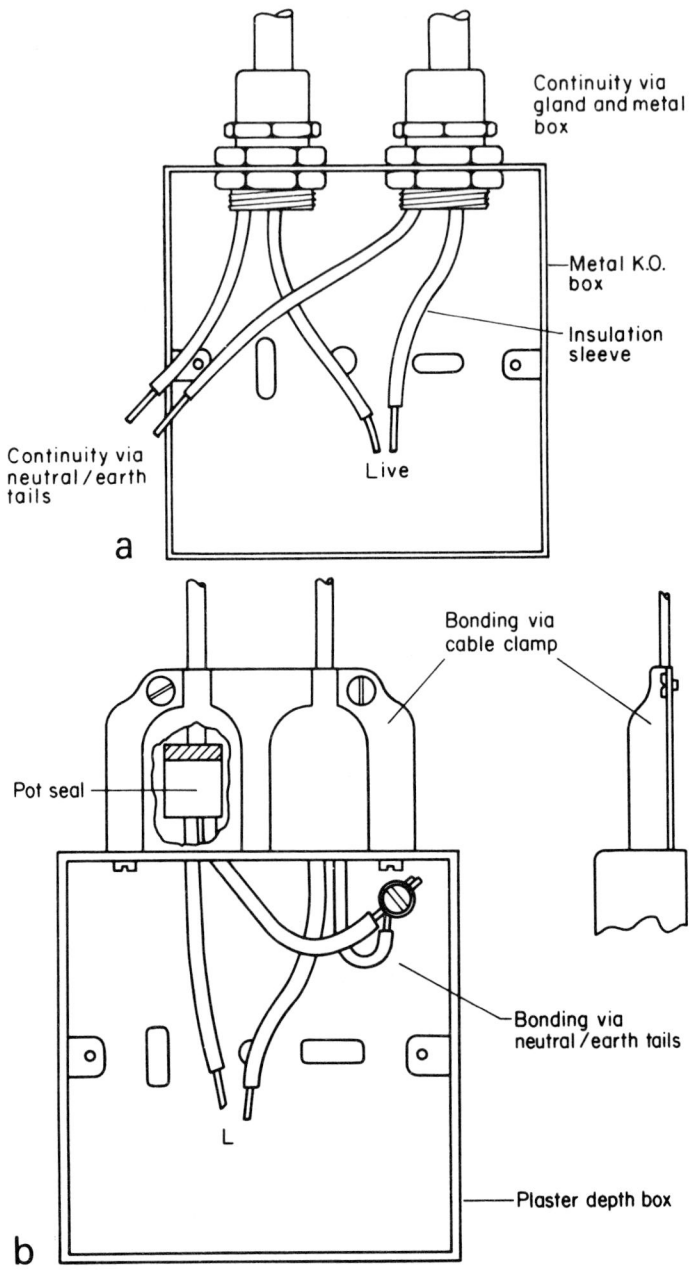

Diagram 97 (a) Ring main (earthed concentric); (b) lighting (earthed concentric)

```
L            L                        L
o--         ---         ----------o--(  )--o
o-- -- -- -- -- -- -- -- -- -- -- -o
N/E   Sheath return
                        Earth bonding
```

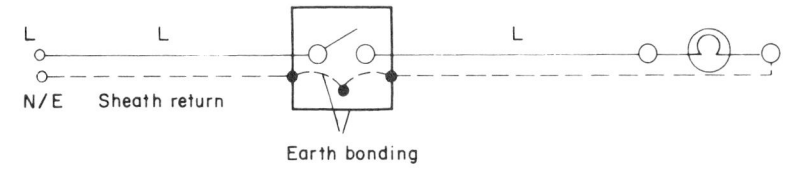

Diagram 98　One-way light switch in earthed concentric wiring

Since the fault is not switched on or off it will be drawing current continuously.

$$\therefore \text{ Energy consumed in 1 year} = \frac{576}{1000} \text{ (kW)} \times 24 \text{ (h)} \times 7 \text{ (days)} \times 52 \text{ (weeks)}$$

$$= 5032 \text{ kWh}$$

$$\text{Cost at 5.7p per unit} = \frac{5032 \times 5.7}{100}$$

$$= £286.82$$

This is clearly an undesirable situation although a not uncommon one, and indicates the desirability of the regular testing of an installation.

Points to note (I.E.E. Regulations)

1. If automatic disconnection of supply is used as the method of protection against indirect contact, then socket outlets forming part of a T.T. system must be protected by a r.c.c.b of 30 mA or less.
2. Earthing leads must be soundly connected to an earth electrode, the connection being readily accessible and labelled 'SAFETY – ELECTRICAL CONNECTION – DO NOT REMOVE', in clear letters of minimum height 4.75 mm.
3. Earthed concentric wiring may be installed only when:
   (a) the supply authority has had permission to allow additional connections of neutral to earth (P.M.E.), or
   (b) if the supply is via a transformer or converter such that there is no metallic connection with a public supply, or
   (c) if the supply is from a private generating plant.
4. The external sheath must be earthed and should not be common to more than one circuit.

5. The earthed external sheath should not contain any fuse, non-linked switch or circuit breaker.
6. There must be *two* bonds across all joints and terminations in the external sheath.

CHAPTER 6

Installation Systems

The I.E.E. Regulations recommend that every consumer's installation should have a means of isolation, a means of overcurrent protection and a means of earth leakage protection. This recommendation applies whatever the size or type of installation, and the sequence of this equipment will be as shown in *Diagrams 99a* and *99b*.

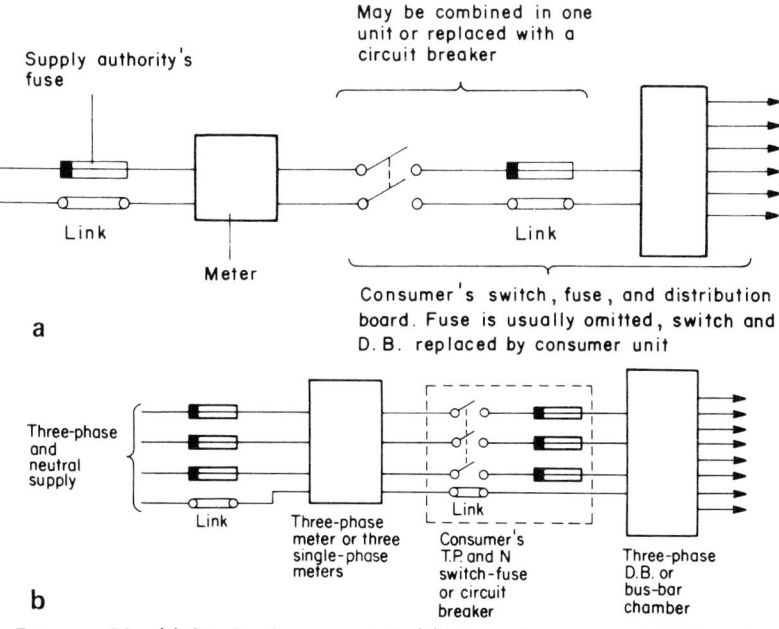

Diagram 99 (a) Single-phase control; (b) three-phase control. T.P. and N = triple pole and neutral

Industrial installations

Industrial installations differ basically from domestic and commercial ones only in the size and type of equipment used. The supplies are three-phase four-wire, and switchgear is usually metal-clad. For extremely heavy loads, switch-fuses are

104 *Installation systems*

replaced by circuit breakers, and much use is made of overhead bus-bar trunking systems. *Diagram 100* shows a typical layout.

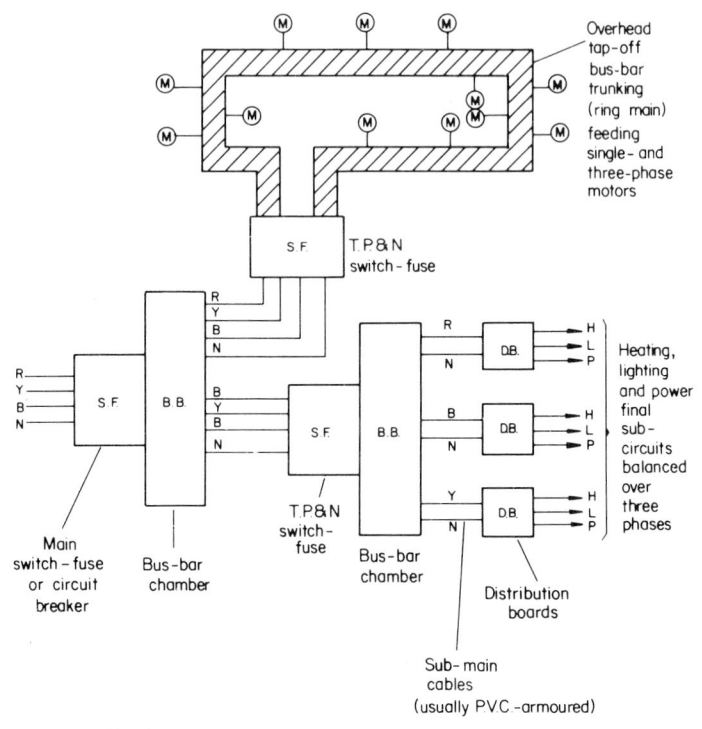

Diagram 100   Layout of industrial installation

Multi-storey commercial or domestic installations

In order to supply each floor or individual flat in a block, it is necessary to run cables from the main intake position. These cables are called *risers*, and the sub-main cables which run from these to each individual supply point are called *laterals*.

The majority of rising mains are in the form of bus-bar trunking with either rectangular or circular conductors; this enables easy tapping off of sub-main cables. *Diagram 101* shows a typical system.

The rising-main system is similar to the ordinary radial circuit in that one cable run supplies several points. Hence the current flowing in the cable at the far end will be less than that at the supply end and the volt drop will be greater at the far end with all loads connected (*Diagrams 102* and *103*).

Installation systems 105

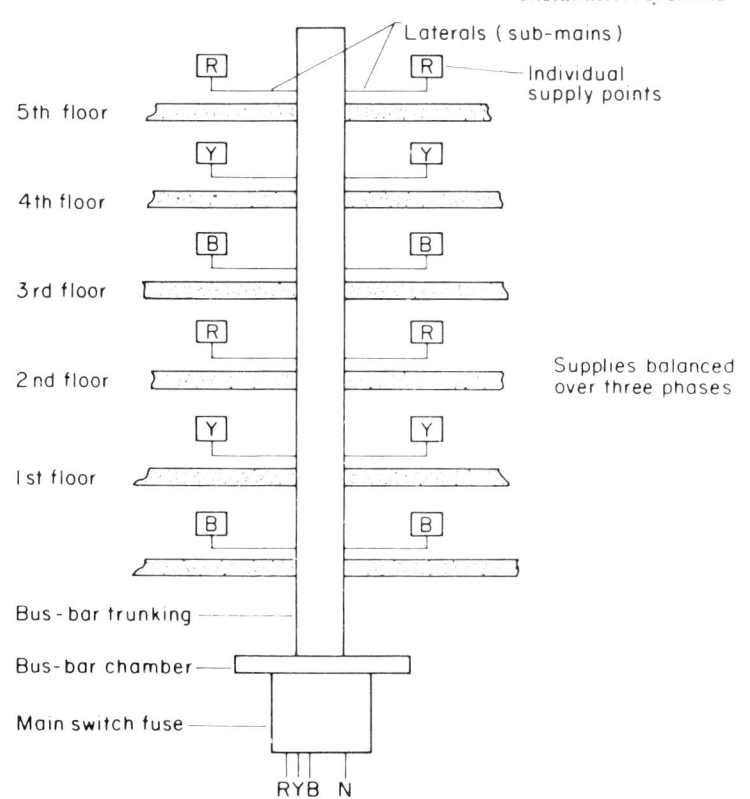

Diagram 101 Rising mains in a block of flats

Diagram 102 Radial circuit

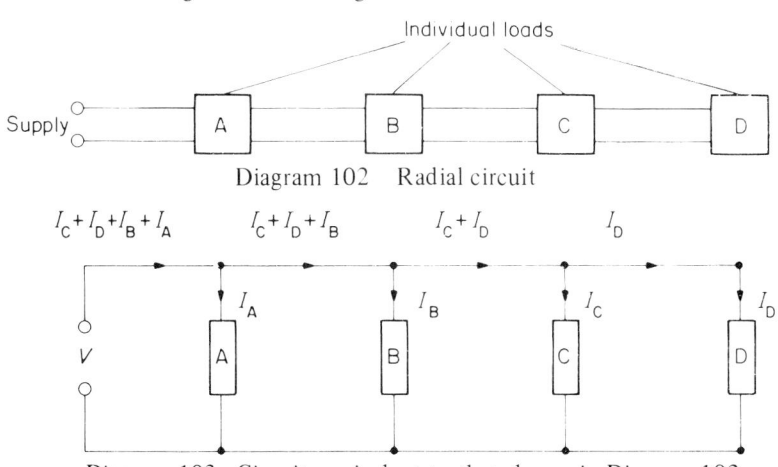

Diagram 103 Circuit equivalent to that shown in Diagram 102

106   *Installation systems*

Hence the currents may be found in any part of a radial distribution cable. Also, if the resistance per metre of the cable is available and the position of the loads along the cable is known, the volt drop at points along the cable may be calculated.

Example 6.1

A 240 V radial distributor is 70 m long and has a resistance of 0.0008 Ω per metre supply and return. Four loads A, B, C and D rated at 30 A, 45 A, 60 A and 80 A are fed from the cable at distances of 20 m, 10 m, 15 m and 25 m respectively. Calculate the total current drawn from the supply, the current in the cable between each of the loads, and the voltage at load D if all the loads are connected (*Diagram 104*).

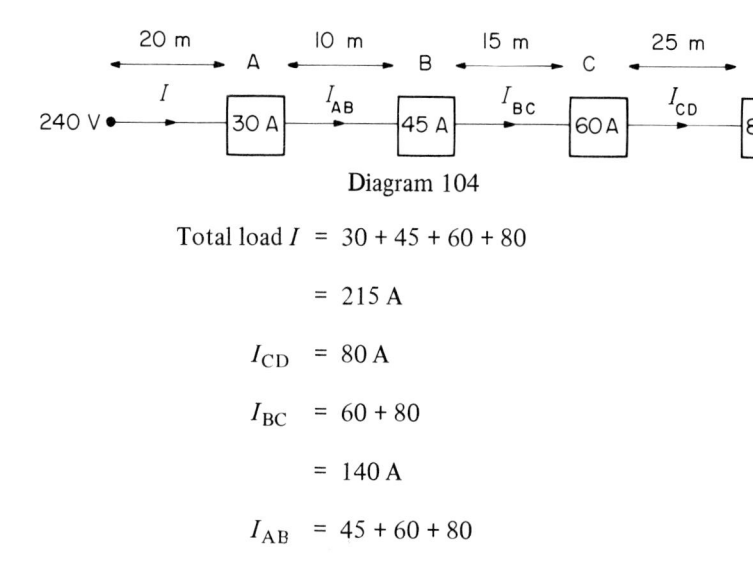

Diagram 104

$$\text{Total load } I = 30 + 45 + 60 + 80$$

$$= 215 \text{ A}$$

$$I_{CD} = 80 \text{ A}$$

$$I_{BC} = 60 + 80$$

$$= 140 \text{ A}$$

$$I_{AB} = 45 + 60 + 80$$

$$= 185 \text{ A}$$

$$\text{Resistance between supply and A} = 20 \times 0.0008$$

$$= 0.016 \text{ Ω}$$

$$\therefore \text{Volt drop between supply and A} = 0.016 \times I$$

$$= 0.016 \times 215$$

$$= 3.44 \text{ V}$$

Resistance between A and B = 10 × 0.0008

= 0.008 Ω

∴ Volt drop between A and B = 0.008 × I_{AB}

= 0.008 × 185

= 1.48 V

Resistance between B and C = 15 × 0.0008

= 0.012 Ω

∴ Volt drop between B and C = 0.012 × I_{BC}

= 0.012 × 140

= 1.68 V

Resistance between C and D = 25 × 0.0008

= 0.02 Ω

∴ Volt drop between C and D = 0.02 × I_{CD}

= 0.02 × 80

= 1.6 V

Total volt drop = 3.44 + 1.48 + 1.68 + 1.6

= 8.2 V **(this is not acceptable)**

∴ Voltage at load D = 240 − 8.2

= 231.8 V

It is interesting to note that if load A were switched off:

Resistance between supply and B = 30 × 0.0008

= 0.024 Ω

108 *Installation systems*

$$\therefore \text{Volt drop between supply and B} = 0.024 \times I_{AB}$$

$$= 0.024 \times 185$$

$$= 4.44$$

$$\therefore \text{Total volt drop now} = 4.44 + 1.68 + 1.6$$

$$= 7.72 \text{ V } \textbf{(still not acceptable)}$$

and

$$\text{Voltage at load D} = 232.28 \text{ V}$$

This example illustrates the need for careful calculation of cable size, length and loading to ensure that no serious loss of voltage occurs at the end of a radial circuit.

Off-peak supplies

As the name implies, off-peak electricity is supplied to the consumer at a time, usually between 11 p.m. and 7 a.m., when demand is not at a peak. This ensures a greater economy in the use of generators and hence the cost per unit to the consumer is low.

These supplies are used mainly for space and water heating; however, some intake arrangements allow all energy-using devices to be used on off-peak.

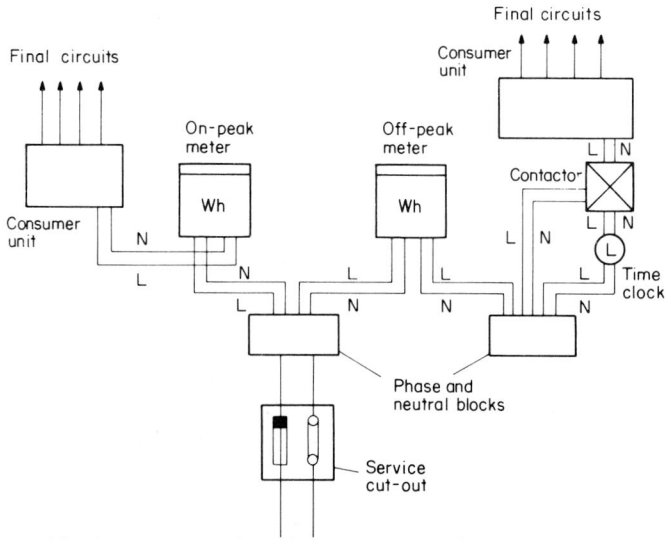

Diagram 105   Arrangement for off-peak supplies (earths omitted for clarity)

Standard off-peak arrangement

Diagram 105 shows the arrangement at the supply intake position. In this arrangement the time clock controls the contactor coil. In installations with only a light off-peak load, the time clock contacts are able to control the load directly.

The white meter

A white meter has two recording dials, one for normal supplies, the other for off-peak. Two time clocks and a contactor are used to change over to off-peak. With this system the whole installation can be run during off-peak hours. *Diagram 106* illustrates this system.

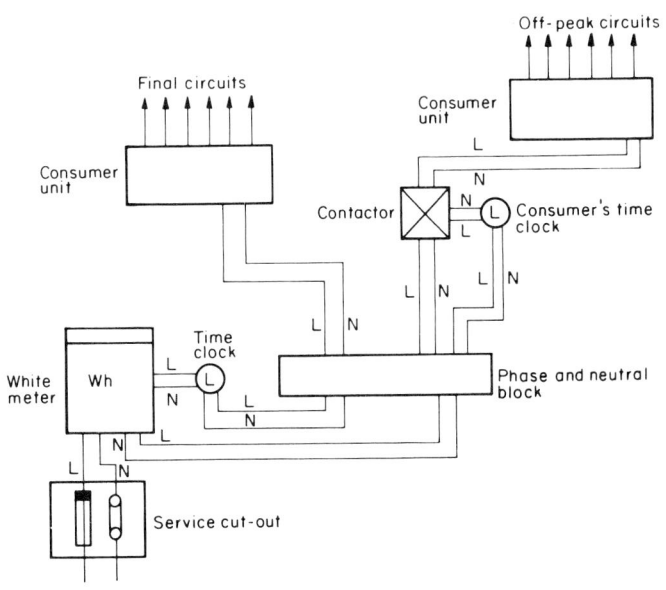

Diagram 106 White meter

During normal hours, only the normal-rate dial will record. The consumer can in fact use his off-peak appliances during this period, by overriding his time clock. Of course, any energy used by these appliances is charged at the normal rate.

At a pre-set time, say 11 p.m., the authority's time clock automatically changes the connections to the dials in the meter, and energy used by any appliance will be metered at off-peak rates.

Choice of system

The choice of any particular wiring system and its accessories will depend on the environment in which it is to be installed. Under normal conditions, typical wiring systems would include:

P.V.C.-insulated P.V.C.-sheathed	– domestic premises; small shops and offices, etc.
P.V.C. conduit or trunking	– offices; light industry
Metal conduit; trunking or armoured cable	– any situation where there is a serious risk of mechanical damage
M.i.m.s.	– fire alarm systems; boilerhouses; earthed concentric wiring, etc.

There are, however, certain environments which need special attention.

Temporary and construction site installations

Installations such as these always present a hazardous situation and special attention must be paid to guarding against mechanical damage, damp and corrosion, and earth leakage.

Any such installation should conform with B.S.C.P. 1017 and be in the charge of a competent person who will ensure that it is safe, and his or her name and designation must be displayed clearly adjacent to the main switch of the installation. The associated distribution board should conform with B.S. 4363.

The installation should be tested at intervals of no more than three months, this period being shown clearly on any completion certificate issued.

As an added precaution against shock risk, portable tools should be supplied from a double-wound transformer with reduced secondary voltage and the secondary winding centre tapped to earth. This ensures that any fault to earth presents only 55 V to the user.

Installations in flammable and/or explosive situations

In premises such as petrol stations, gas works, flour mills, etc., using electricity involves an obvious risk. Serious arcing at contacts or in faulty conductors or equipment could cause an explosion and/or fire. It is therefore important to ensure that such a dangerous situation does not arise, by installing suitable cable and fittings. An outline of the hazards and suitable wiring systems is given in C.P. 1003. The following paragraphs are extracts from that Code of Practice, reproduced by permission of the B.S.I., 2 Park Street, London W1A 2BS, from whom complete copies of the Code of Practice may be obtained.

Extracts from B.S. Code of Practice 1003 Part 1, 1964: 'Electrical apparatus and associated equipment for use in explosive atmospheres of gas or vapour (other than mining)'. Part 1: 'Choice, installation and maintenance of flameproof and intrinsically safe equipment'

Foreword

In dealing with the risk of fire or explosion from the presence of flammable liquids, gases or vapours, three sets of conditions are recognized:

Division 0

An area or enclosed space within which any flammable or explosive substance, whether gas, vapour or volatile liquid, is continuously present in concentrations within the lower and upper limits of flammability.

Division 1

An area within which any flammable or explosive substance, whether gas, vapour or volatile liquid, is processed, handled or stored, and where during normal operations an explosive or ignitable concentration is likely to occur in sufficient quantity to produce a hazard.

Division 2

An area within which any flammable or explosive substance, whether gas, vapour or volatile liquid, although processed or stored, is so well under conditions of control that the production (or release) of an explosive or ignitable concentration in sufficient quantity to constitute a hazard is only likely under abnormal conditions.

The conditions described as appertaining to Division 0 are such as normally to require the total exclusion of any electrical equipment, except in such special circumstances as to render this impracticable, in which case recourse may be possible to special measures such as pressurization or the use of intrinsically safe equipment.

A risk of the nature described under Division 1 can be met by the use of flameproof or intrinsically safe equipment, with which Part 1 is concerned, or by the use of the means described in Part 2 of the Code which is concerned with methods, other than the use of flameproof or intrinsically safe equipment, of securing safety in flammable and explosive atmospheres.

The certifying authority for flameproof apparatus is the Ministry of Power. The certifying authority for intrinsically safe electrical apparatus for use in factories coming within the scope of the Factories Act, is the Ministry of Labour. The recognized testing authority for flameproof enclosures and intrinsically safe circuits and apparatus is the Ministry of Power*.

---

*The Ministry of Power is now called the Department of Energy, and the Factories Act is now overseen by the Health and Safety Executive.

Types of hazard

Two main types to be considered: gases and vapours or flammable liquids.

(a) *Explosive gases* or vapours are grouped according to the grade of risk and four groups are recognized.

- Group I. Gases encountered in coal mining.
- Group II. Various gases commonly met with in industry.
- Group III. Ethylene, diethyl ether, ethylene oxide, town gas and coke-oven gas.
- Group IV. Acetylene, carbon disulphide, ethyl nitrate, hydrogen and water-gas.

(i) Flameproof apparatus. Apparatus with flameproof enclosures, certified appropriately for the gas group which constitutes the risk, should be used. It should be noted that no apparatus is certified for Group IV; other techniques, e.g. pressurization, must therefore be applied if electrical apparatus has to be installed where gases in this group may be present in dangerous concentrations.

In general, apparatus certified for the higher groups will cover situations where gases from the lower groups are present.

(ii) Selection of apparatus for diverse risks. If, in an installation capable of subdivision, Group III apparatus is required for some parts while Group II would suffice for other parts, it is recommended that the former, which could cover the risks in a lower Group, should be used throughout lest apparatus of the latter group should inadvertently be transferred to a place where Group III is required.

(b) *Flammable liquids*. Flammable liquids give rise, in a greater or lesser degree according to their flashpoints and the temperature to which they are subjected, to explosive vapours which should be treated as under (a) above.

The liquid, however, constitutes a further risk in that fires may occur as a result of unvaporized liquid in the form of spray or otherwise, coming into contact with electrical equipment and then being ignited by a spark or other agency.

Types of wiring

Danger areas:

(a) Cables drawn into screwed solid-drawn steel conduit.
(b) Lead-sheathed, steel-armoured cable.
(c) Mineral-insulated, metal-sheathed cable.
(d) P.V.C.-insulated and armoured cable with an outer sheath of P.V.C.
(e) Polyethylene-insulated, P.V.C.-covered overall and armoured.
(f) Cables enclosed in a seamless aluminium sheath with or without armour.

Automatic electrical protection

(a) All circuits and apparatus within a danger area should be adequately protected against overcurrent, short circuit and earth leakage current.

(b) Circuit breakers should be of the free-handle trip-free type to preclude misuse, such as tying-in or holding-in under fault conditions against the persistence of which they are designed to afford protection, and an indicator should be provided in all cases to show clearly whether the circuit breaker is open or closed.

Portable and transportable apparatus and its connections

Portable electrical apparatus should only be permitted in any hazardous area in the most exceptional circumstances which make any other alternative extremely impracticable, and then only if it is of a certified type.

Agricultural and horticultural installations

Any sort of farm or smallholding requiring the use of electricity presents hazards due to dampness, corrosion, mechanical damage, shock risk and fire/explosion risk.

It is therefore important to install wiring systems and apparatus that will reduce these hazards.

Switchgear

Main switchgear should be accessible at all times even in the event of livestock panic and should be located out of reach of livestock.

Most farms and smallholdings have several buildings and/or glass houses fed from the installation, and each building should therefore have its own control switchgear located in or adjacent to it.

Any items of equipment remote from the main installation should also have control switchgear located close by.

It is very likely with large installations that both three-phase and single-phase circuits will exist, and switchgear and distribution board covers should be marked to show the voltage present.

All points including socket outlets must be switched and switches controlling machinery should have the ON and OFF positions clearly marked.

Cables

Cables are the parts of an installation most susceptible to damage and great care should be taken to choose and position them correctly.

All cables should be kept out of the reach of livestock and clear of vehicles. If a long run is to be placed along the side of a building it should, if possible, be run on the outside and as high as is practicable.

Ideally, only non-metallic conduit should be used, but when the use of metal tubing is unavoidable, only heavy-gauge galvanized conduit should be installed.

114 *Installation systems*

When it is necessary to wire between buildings, the cable may be underground at a depth sufficient to avoid damage by farm implements, or overhead, supported by buildings or poles, but not housed in steel conduit or pipes.

P.V.C.- or h.o.f.r.-sheathed cables (h.o.f.r. stands for heat-resisting, oil-resisting and flame-retardant) should not be installed where there is a chance of contact with liquid creosote, as this substance will damage the sheathing. Rubber-sheathed cables should be used indoors only in clean, dry situations.

Non-sheathed flexible cords of the twisted or parallel twin types must not be used, nor must cable couplers.

Earthing

In environments such as farms an 'all-insulated' type of installation is preferable. However, if this is not practicable the main regulations on earthing must be strictly observed. In addition, no metallic conduit must be used as the sole C.P.C., and earthing conductors must be protected against damage.

Electric fence.

Mains-operated fence controllers can be a source of danger as there is always a chance that the output could, under fault conditions, be connected to the low-voltage supply. Hence they should not be placed where there is likelihood of mechanical damage or interference by unauthorized persons. They should not be fixed to poles carrying power or telecommunication lines. They can, however, be fixed to a pole used solely for carrying an insulated supply to themselves.

Any earth electrode associated with a controller must be outside the resistance area of any electrode used for normal protective earthing, and only one controller should be used for a fence system.

When a battery-operated fence controller is being charged, the battery must be disconnected from the controller.

General

In some agricultural situations such as grain-drying areas, there is a risk of explosion from flammable dusts and in such situations all apparatus should be selected very carefully to prevent dangerous conditions from arising.

Edison-type screw lampholders must be of the drip-proof type and persons using wash troughs and sterilizing equipment, etc. must not be able to come into contact with switchgear or non-earth-bonded metalwork.

Finally, testing should be carried out at least once every three years.

Cable selection

Once the type and system of wiring to be installed has been decided, it is necessary to calculate the design current I_B of the individual circuits. The value of protection I_N is then chosen such that $I_N \geq I_B$. Then, depending on the environmental conditions and type of protection, the current-carrying capacity of the cable may be determined.

Installation systems 115

This capacity is adjusted using certain correction factors listed in the Regulations. The application of these factors ensures that the cable insulation will not be damaged in adverse environmental conditions.

Example 6.2

A single-phase fixed load of 9.6 kW is to be fed at 240 V by a P.V.C.-sheathed non-armoured twin with earth copper cable 15 m long. The cable is clipped surface through an area whose ambient temperature is 45°C, and is grouped with two other cables of similar size and loading. The protection is by means of a type 2 B.S. 3871. Calculate the minimum cable size and the volt drop, if the external loop impedance is 0.4 Ω.

$$\text{Design current } I_B = \frac{P}{V}$$

$$= \frac{9600}{240}$$

$$I_B = 40 \text{ A}$$

$$\therefore \text{ size of CB} = 50 \text{ A} = I_N$$

Correction factors from tables:

For ambient temperature = 0.79
For groups of cables (3) = 0.7

These may be combined for an overall correction factor which will be

$$\text{Overall C.F.} = 0.79 \times 0.7 = 0.553$$

$$\therefore \text{current-carrying capacity of cable} = \frac{\text{rating of protection } I_N}{\text{overall C.F.}}$$

$$= \frac{50}{0.553}$$

$$= 90.4$$

Of course this current will never flow; the conditions in whch the cable is run simply produce the effect of 90.4 A. Hence a cable must be selected to cater for this value.

From the tables, a 25 mm² cable will carry 108 A. It will therefore carry the actual load current safely.

Check for volt drop: The millivolt drop per ampere (load current) per metre run of 16 mm² cable from the tables is 1.8 mV.

116 Installation systems

$$\therefore \text{Volt drop along cable} = \frac{1.8 \times 40 \times 15}{1000}$$

$$= 1.08 \text{ V}$$

This is acceptable since the maximum permissible volt drop = 2.5% of 240 V = 6 V,

$$\therefore \text{Cable size is } 25 \text{ mm}^2$$

It is also important to ensure that the C.P.C. is of the correct size, for shock risk and thermal constraints.

Shock risk

It is important that, in the event of a fault to earth, the circuit protection operates before there is a risk of electric shock.

For socket outlet circuits and bathrooms the disconnection time should not exceed 0.4 seconds; for circuits feeding fixed equipment, it should not exceed 5 seconds.

These disconnection times will be dependent on the loop impedance, maximum values of which are given in tables 41a1 and 41a2, I.E.E. Regulations.

$$Z_S = Z_E + R_1 + R_2$$

where Z_E = External loop impedance
R_1 = Resistance of phase conductor
R_2 = Resistance of C.P.C.

This value of Z_S must not exceed the tabulated maximum value.

For a Type 2 50 A M.C.B. feeding fixed equipment, the maximum value of Z_S from table 41a2 is 0.68 Ω.

A 25 mm² twin cable has a 10 mm² C.P.C.

$$\therefore R_1 + R_2 = 2.557 \text{ m}\Omega/\text{m (from table 17A, I.E.E. Regulations)}$$

$$\therefore \text{ for a 15 m length, } R_1 + R_2 = \frac{2.557 \times 15}{1000} = 0.038 \text{ }\Omega$$

However, this value is at 20°C, so to find the value under fault conditions, when the cable will be hot, table 17B is used:

$$\therefore \text{actual } R_1 + R_2 = 0.038 \times 1.38 = 0.052$$

$$\therefore \text{actual } Z_S = Z_E + R_1 + R_2$$

$$= 0.4 + 0.052$$

$$= 0.452 \text{ }\Omega$$

As this is less than the 0.68 Ω maximum, shock risk protection is assured.

Thermal constraints

Under fault conditions, large currents flow and a check must be made to ensure that the C.P.C. can carry such currents without damage to insulation.

The adiabatic equation is used to check this:

$$S = \frac{\sqrt{I_f^2 \, t}}{k}$$

where S = Minimum C.S.A. of C.P.C.
I_f = Fault current (from $I_f = (U_0/Z_S)$)
t = Disconnection time from relevant curve
k = Factor from tables 54B to 54E, I.E.E. Regulations

$$\therefore I_f = \frac{U_0}{Z_S} = \frac{240}{0.452} = 531 \text{ A}$$

$t = 0.03$ secs (curve on fig. 14, I.E.E. Regulations)

$k = 115$ (table 54C)

$$\therefore \text{C.P.C. size} = S = \frac{\sqrt{I_f^2 \, t}}{k} = \frac{\sqrt{531^2 \times 0.03}}{115} = 0.8 \text{ mm}^2$$

We have a 10 mm² C.P.C. which is clearly acceptable.

Example 6.3

A 12.6 kW, three-phase 415 V delta-connected motor is 86% efficient and runs at a P.F. of 0.9 lagging. It is to be supplied by a 29 m length of three-core mineral-insulated copper-clad P.V.C.-covered cable run in an ambient temperature of 40°C. Calculate the minimum permissible cable size and the volt drop. The protection is given by high-breaking-capacity (H. B. C.) fuses to B.S. 88 part 2.

Motor ratings in kilowatts refer to output, therefore

$$\text{Efficiency (\%)} = \frac{\text{output}}{\text{input}} \times 100$$

$$\text{Input kW} = \frac{\text{output kW}}{\text{efficiency}} \times 100$$

$$= \frac{12.6 \times 100}{86}$$

$$= 14.65 \text{ kW}$$

$$\therefore \text{kVA} = \frac{\text{kW}}{\text{P.F.}}$$

118 *Installation systems*

$$= \frac{14.65}{0.9}$$

$$= 16.28 \text{ kVA}$$

$$\therefore \text{ Line current } = \frac{VA}{V_L \sqrt{3}}$$

$$= \frac{16\,280}{415 \times \sqrt{3}}$$

$$= 22.65 \text{ A}$$

This is the design current I_B

$$\therefore \text{ Fuse rating } I_N \text{ will be 25 A}$$

Correction factor:

$$\text{Ambient temperature} = 0.94$$

$$\therefore \text{ Current-carrying capacity of cable } = \frac{25}{0.94}$$

$$= 26.6 \text{ A}$$

$$\therefore \text{ Cable size to suit will be } 4.0 \text{ mm}^2$$

Check for volt drop:

$$\text{Volt drop} = \frac{9 \times 22.65 \times 29}{1000}$$

$$= 5.9 \text{ V}$$

This is acceptable, as the maximum permissible volt drop is

$$415 \times 2.5\% = 10.375 \text{ V}$$

$$\therefore \text{ Cable size is } 4.0 \text{ mm}^2$$

Once again, before this size of cable can be used, a check must be made to ensure that the C.P.C. is adequate.

In this case, the C.P.C. is the cable sheath and from manufacturers information it will be found that a 3 core 4.0 mm² m.i.c.s. cable has a sheath c.s.a. in excess of 15 mm², which is much larger than the phase conductor and hence no further calculation is necessary.

Questions on Chapter 6

1. A radial distributor is 100 m long and has three loads A, B and C of 30 A, 20 A and 50 A taken from it at 15 m, 70 m and 100 m respectively, from the supply end. Calculate the voltage at load C, if the total cable resistance is 0.0012 Ω/m.

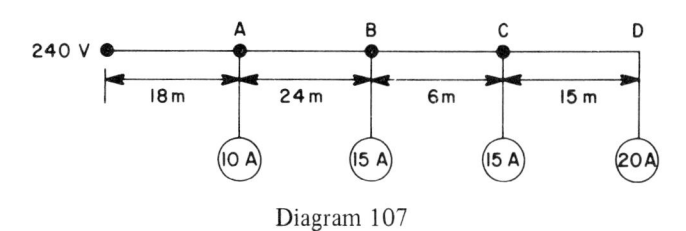

Diagram 107

2. In *Diagram 107*, the cable resistance (L & N) is 0.0026 Ω/m. Calculate the voltage available at D when all loads are connected and when loads B and C are switched off.

3. A 240 V single-phase 12 kW load has a P.F. of 0.88 lagging and is to be supplied by a P.V.C.-sheathed armoured twin cable with copper conductors. The cable is to be installed on cable tray through an area where the ambient temperature is 25°C. The length of the run is 40 m and the protection is afforded by means of a type 2 C.B. to B.S. 3871. Calculate the minimum cable size permissible.

4. A 415 V three-phase load of 30 kW works at a P.F. of 0.9 lagging and is located some 25 m from the distribution board. The cable required to supply the load is to be P.V.C.-armoured copper and run with three other cables through an area where the ambient temperature is 40°C. The protection is to be by means of H.R.C. fuses. Calculate the minimum cable size to allow for a 1.5 V drop between the intake position and the distribution board.

CHAPTER 7
Lighting and Illumination

It will be best to introduce the subject of illumination by listing the various units and quantities used.

Luminous intensity: symbol, I; unit, candela (cd)

This is a measure of the power of a light source and is sometimes referred to as brightness.

Luminous flux: symbol, F; unit, lumen (lm)

This is a measure of the flow or amount of light emitted from a source.

Illuminance: symbol, E; unit, lux (lx) or lumen/m^2

This is a measure of the amount of light falling on a surface. It is also referred to as *illumination*.

Luminous efficacy: symbol, K; unit, lumen per watt (lm/W)

This is the ratio of luminous flux to electrical power input. It could be thought of as the 'efficiency' of the light source.

Maintenance factor (M.F.): no units

In order to allow for the collection of dirt on a lamp and also ageing, both of which cause loss of light, a maintenance factor is used.
    An an example, consider a new 80 W fluorescent lamp with a lumen output of 5700 lm. After about three or four months this output would have fallen and settled at around 5200 lm. Hence the light output has decreased by

$$\frac{5200}{5700} = 0.9$$

    This value, 0.9, is the maintenance factor and should not fall below 0.8. This is ensured by regular cleaning of the lamps.

Lighting and illumination 121

Coefficient of utilization (C.U.): no units

The amount of useful light reaching a working plane will depend on the lamp output, the reflectors and/or diffusers used, position of lamp, colour of walls and ceilings, etc. The lighting designer will combine all of these considerations and determine a figure to use in his or her lighting calculations.

Inverse-square law

If we were to illuminate a surface by means of a lamp positioned vertically above it, measure the illumination at the surface, and then move the lamp twice as far away, the illumination now measured would be four times less. If it were moved away three times the original distance the illumination would be nine times less. Hence it will be seen that the illuminance on a surface is governed by the square of the vertical distance of the source from the surface (*Diagram 108*).

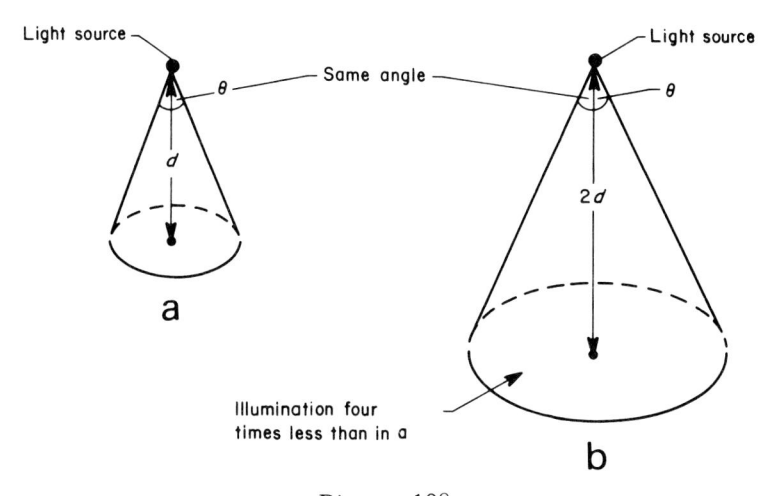

Diagram 108

$$\therefore \text{Illuminance } E \text{ (lux)} = \frac{\text{luminous intensity (candelas)}}{d^2}$$

$$E = \frac{I}{d^2}$$

Example 7.1

A light source of 900 candelas is situated 3 m above a working surface. (a) Calculate the illuminance directly below the source. (b) What would be the illuminance if the lamp were moved to a position 4 m from the surface?

122 *Lighting and illumination*

$$E = \frac{I}{d^2}$$

$$= \frac{900}{9}$$

$$= 100 \text{ lx}$$

$$E = \frac{I}{d^2}$$

$$= \frac{900}{16}$$

$$= 56.25 \text{ lx}$$

Cosine rule

From *Diagram 109*, it will be seen that point X is further from the source than is point Y. The illuminance at this point is therefore less. In fact the illuminance at X depends on the cosine of the angle θ, hence

$$E_X = \frac{I \times \cos^3 \theta}{d^2}$$

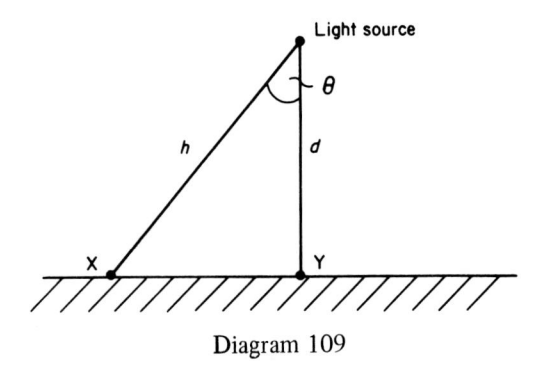

Diagram 109

Example 7.2

A 250 W sodium vapour street lamp emits a light of 22 500 cd and is situated 5 m above the road. Calculate the illuminance (a) directly below the lamp and (b) at a horizontal distance along the road of 6 m.

Lighting and illumination 123

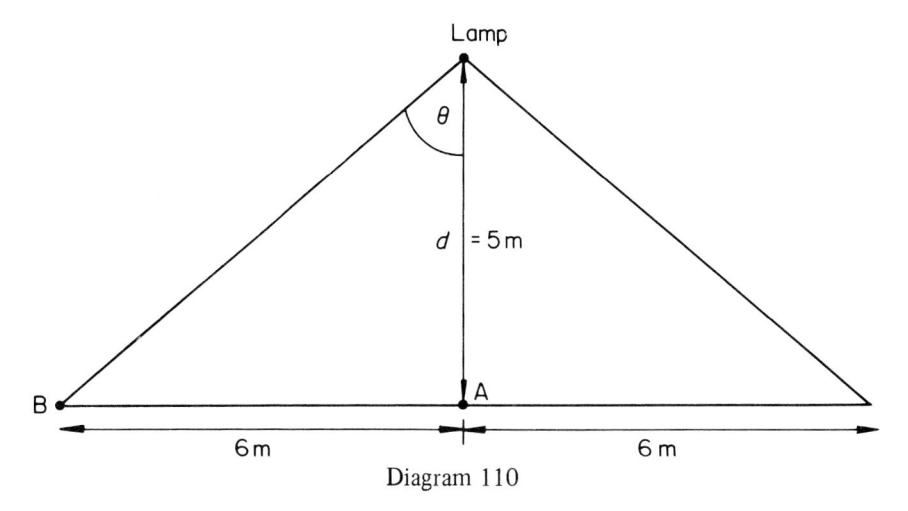

```
                    Lamp
                     ↑
                    /|\
                   / θ \
                  /  |  \
                 /   |   \
                /  d |=5m \
               /     |     \
              /      |      \
             /       |       \
            /        |A       \
        B •----------•----------•
              6 m         6 m
```
Diagram 110

From *Diagram 110*, it can be seen that the illuminance at A is given by

$$E_A = \frac{I}{d^2}$$

$$= \frac{22\,500}{25}$$

$$= 900\,\text{lx}$$

The illuminance at B is calculated as follows: Since the angle θ is not known, it can be found most simply by trigonometry.

$$\tan\theta = \frac{AB}{d}$$

$$= \frac{6}{5} = 1.2$$

From tangent tables,

$$\theta = 55.77°$$

and from cosine tables,

$$\cos 55.77° = 0.64$$

$$\therefore E_B = \frac{I\cos^3\theta}{d^2}$$

$$= \frac{22\,500 \times 0.64^3}{25}$$

$$= 236 \text{ lx}$$

Calculation of lighting requirements

In order to estimate the number and type of light fittings required to suit a particular environment, it is necessary to know what level of illuminance is required, the area to be illuminated, the maintenance factor and the coefficient of utilization, and the efficacy of the lamps to be used.

Example 7.3

A work area at bench level is to be illuminated to a value of 300 lx, using 85 W single fluorescent fittings having an efficacy of 80 lumens/watt.

The work area is 10 m × 8 m, the M.F. is 0.9 and the C.U. is 0.6. Calculate the number of fittings required.

$$\text{Total lumens } (F) \text{ required} = \frac{E(\text{lx}) \times \text{area}}{\text{M.F.} \times \text{C.U.}}$$

$$F = \frac{300 \times 10 \times 8}{0.8 \times 0.6}$$

$$= 50\,000 \text{ lm}$$

Since the efficacy is 80 lm/W,

$$\text{Total power required} = \frac{50\,000}{80}$$

$$= 625 \text{ W}$$

As each lamp is 85 W,

$$\text{Number of lamps} = \frac{625}{85}$$

$$= 8$$

Light sources

The range of modern lighting fittings and lamps is so large that only the basic types will be considered here.

Tungsten filament lamp

Diagram 111 shows the basic components of a tungsten filament lamp.

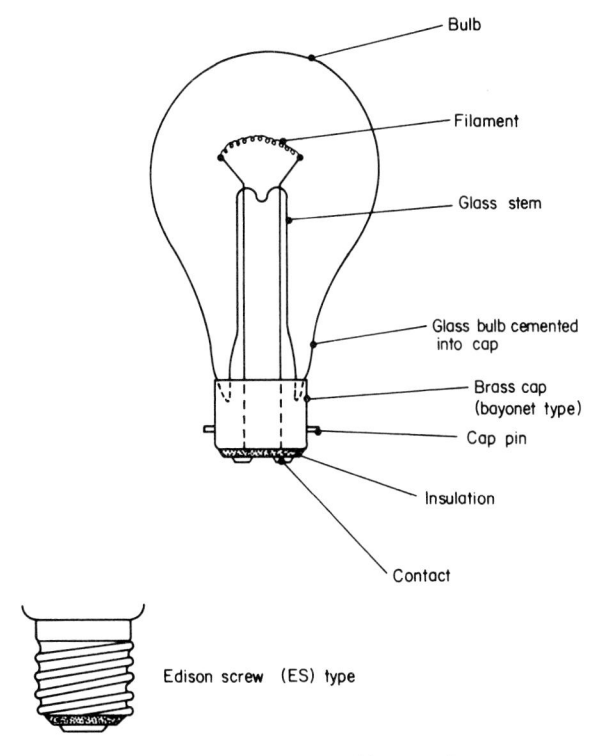

Diagram 111   Tungsten filament lamp

The tungsten filament is either single- or double-coiled (coiled-coil). *Diagram 112* illustrates these types.

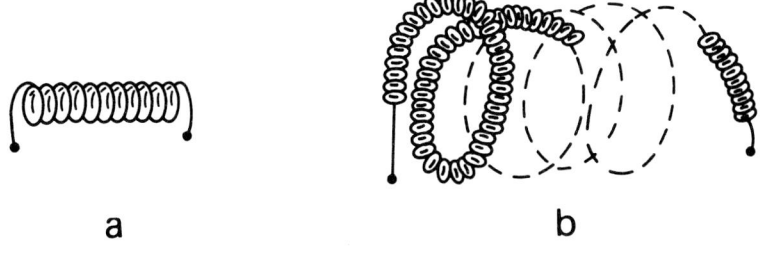

Diagram 112   Filament arrangements. (a) Single coil; (b) coiled coil

The efficacy of gas-filled lamps is increased by using a coiled-coil filament, as this type has in effect a thicker filament which reduces the heat loss due to convection currents in the gas.

Filament lamps are of two main types: vacuum and gas-filled.

Vacuum type

The filament operates in a vacuum in the glass bulb. It has a poor efficacy as it can operate only up to around $2000°C$.

Gas-filled type

In this case the bulb is filled with an inert gas such as nitrogen or argon. This enables the operating temperature to reach $2500°C$. The efficacy increases and the bulb is usually so bright that it is given an opaque coating internally. This type of lamp is usually called a 'pearl' lamp.

The following code refers to lamp caps:

B.C.	–	Bayonet
S.B.C.	–	Small bayonet
S.C.C.	–	Small centre contact
E.S.	–	Edison screw
S.E.S.	–	Small Edison screw
M.E.S.	–	Miniature Edison screw
G.E.S.	–	Goliath Edison screw

The efficacy of a tungsten lamp will depend on several factors including the age of the lamp and its size, but tends to be around 12 lm/W for a 100 W lamp.

The colour of its light tends to be mostly red and yellow and in its basic form this type of lamp is used only in situations that do not require a high level of illumination.

Other lamps of the filament type include tubular strip lights, oven lamps, infra-red heating lamps, spot- and floodlights, and tungsten–halogen lamps.

Discharge lighting

This type of lighting relies on the ionization of a gas to produce light. As high voltages are present in such lighting circuits, special precautions, outlined in the I.E.E. Regulations, must be taken. Typical discharge lamps include decorative neon signs, fluorescent lighting, and mercury and sodium vapour lamps used for street lighting.

Neon tube

In the same way that the trade name 'Hoover' is colloquially used to indicate

Lighting and illumination 127

any make of vacuum cleaner, so 'neon' tends to be used to describe any sort of gas-filled tube. There are in fact several different gases used to give different colours, including helium, nitrogen and carbon dioxide.

Diagram 113 shows the basic circuit for a cold-cathode neon sign installation.

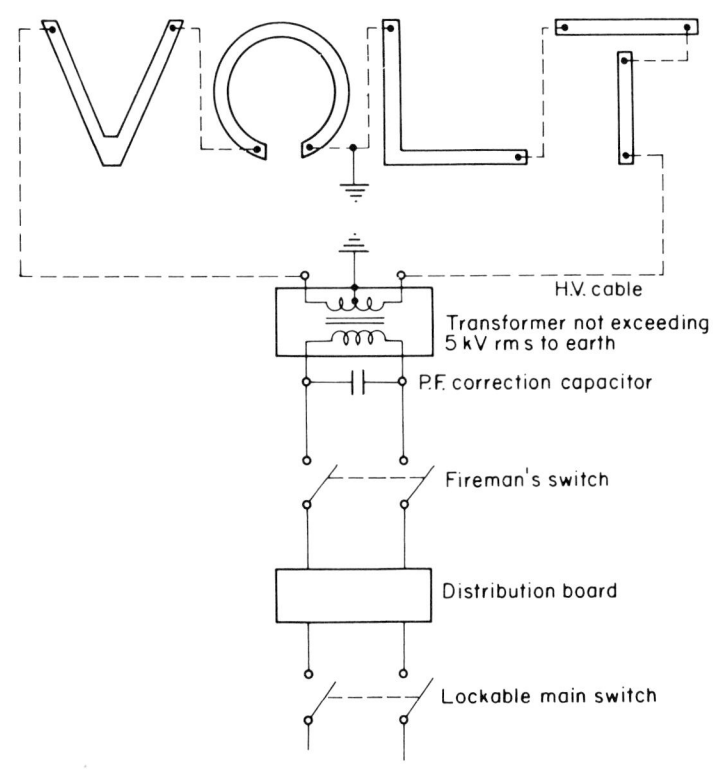

Diagram 113 Simple neon sign circuit

Sodium vapour lamp

There are two types of sodium vapour lamp available, working at high pressure and low pressure respectively.

The low-pressure type consists of a U-shaped double-thickness glass tube, the inner wall of which is of low-silica glass which can withstand attack by hot sodium. Inside the tube is a quantity of solid sodium and a small amount of neon gas (this helps to start the discharge process). An outer glass envelope stops too much heat loss from the inner tube. *Diagram 114* shows the components of a low-pressure sodium vapour lamp, while *Diagram 115* shows the control circuit for a sodium vapour lamp.

128　*Lighting and illumination*

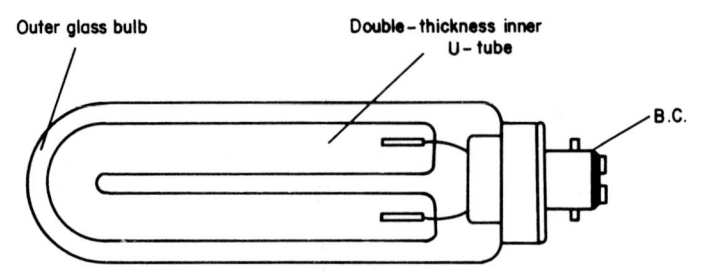

Diagram 114　Low-pressure sodium vapour lamp type SOX, B.C. = bayonet cap

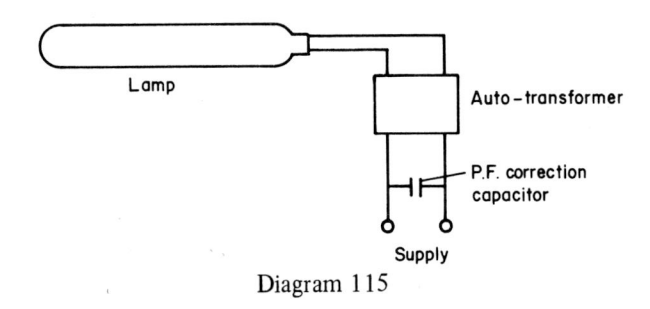

Diagram 115

　　　The output from the auto-transformer is in the region of 480 V and the P.F. correction capacitor is important, as the P.F. of the lamp and transformer can be as low as 0.3 lagging.
　　　The recommended burning position of the lamp is horizontal ± 20°; this ensures that hot sodium does not collect at one end of the tube in sufficient quantities to attack and damage it.
　　　The light output is almost pure yellow, which distorts surrounding colours, and as such is useful only for street lighting. The modern SOX type (superseding the SOH type) has a high efficacy, a 90 W lamp giving in the region of 140 lm/W. (The SOH type gives around 70 lm/W.)
　　　The high-pressure type of sodium vapour lamp differs from other discharge lamps in that the discharge tube is made of compressed aluminium oxide, which is capable of withstanding the intense chemical activity of the sodium vapour at high temperature and pressure. The efficacy is in the region of 100 lm/W, and the lamp may be mounted in any position. The colour is a golden white and as there is little surrounding colour distortion, it is suitable for many applications including shopping centres, car parks, sports grounds and dockyards.

Disposal instructions

Sodium lamps contain a small quantity of sodium metal, a substance which develops heat in contact with moisture. Old lamps should be disposed of by cracking the outer bulb carefully, close to the cap to let in air slowly. The inners should be placed in a dry container, broken into small pieces, and covered with

Lighting and illumination 129

water from a distance using a hose. After a few minutes the contents may be disposed of in the normal way.

High-pressure mercury vapour lamp

This type consists of a quartz tube containing mercury at high pressure and a little argon gas to assist starting. There are three electrodes, two main and one auxiliary; the latter is used for starting the discharge (*Diagram 116*).

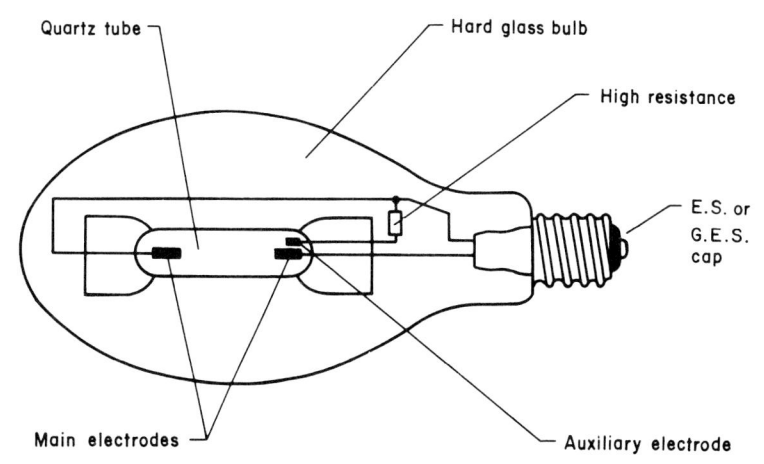

Diagram 116   High-pressure mercury vapour lamp. E.S. = Edison screw type; G.E.S. = goliath Edison screw type

Diagram 117 shows the control circuit for a high-pressure mercury vapour lamp. The initial discharge takes place in the argon gas between the auxiliary electrode and the main electrode close to it. This causes the main electrode to heat up and the main discharge between the two main electrodes takes place.

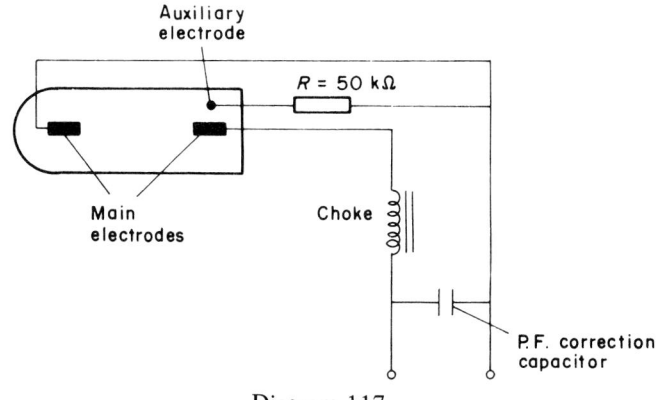

Diagram 117

Several types of mercury vapour lamp are available, including the following two popular types:

M.B. type — Standard mercury vapour lamp; E.S. or G.E.S. cap; any mounting position; efficacy around 40 lm/W. Largely superseded by the M.B.F. type.

M.B.F. type — Standard, but with fluorescent phosphor coating on the inside of the hard glass bulb; E.S. or G.E.S. cap; efficacy around 50 lm/W. Used for industrial and street lighting, commercial and display lighting. Any mounting position.

The colour given by high-pressure mercury vapour lamps tends to be blue-green.

Low-pressure mercury vapour lamp

A low-pressure mercury vapour lamp, more popularly known as a *fluorescent lamp*, consists of a glass tube, the interior of which is coated in fluorescent phosphor. The tube is filled with mercury vapour at low pressure and a little argon to assist starting. At each end of the tube is situated an oxide-coated filament. Discharge takes place when a high voltage is applied across the ends of the tube. *Diagram 118* shows the circuit diagram for a single fluorescent tube.

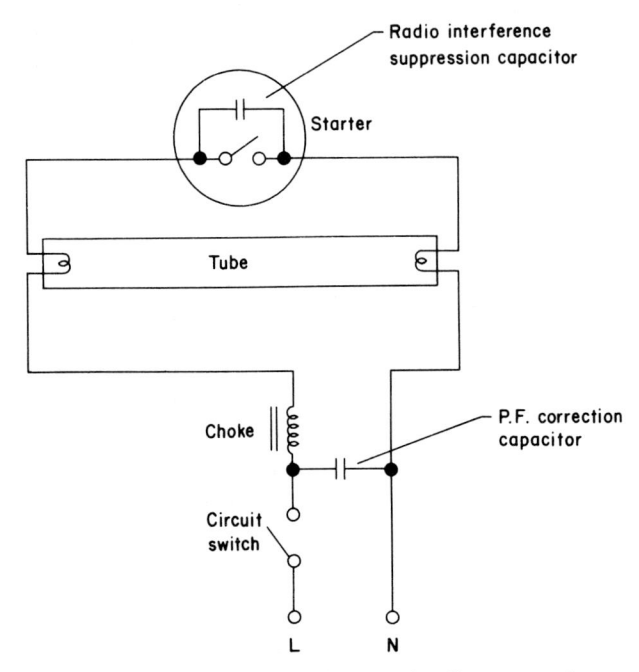

Diagram 118 Basic circuit diagram for fluorescent lamp

Lighting and illumination 131

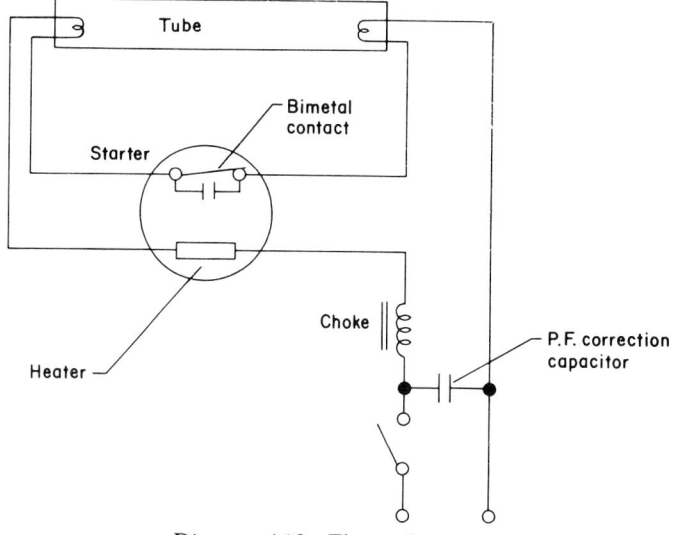

Diagram 119 Thermal starter

Starters

Three methods are commonly available for starting the discharge in a fluorescent tube: the thermal start, the glow start and the quick start.

A *thermal starter* consists of two contacts (one of which is a bimetal) and a heater. *Diagram 119* shows how such a starter is connected.

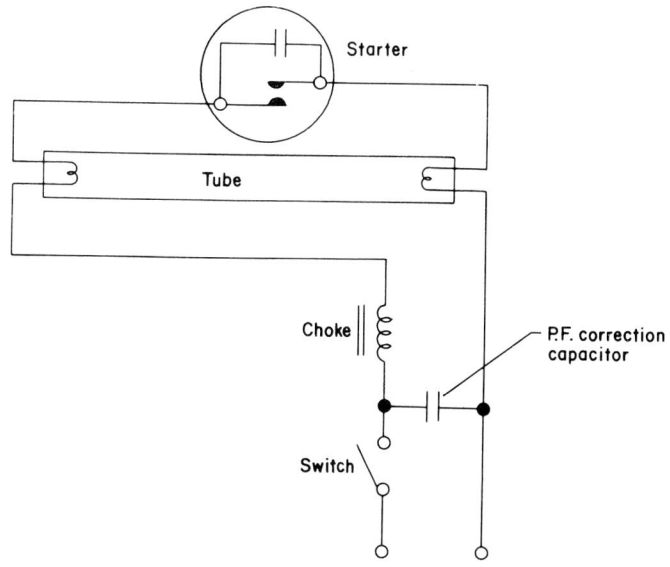

Diagram 120 Glow starter

132   *Lighting and illumination*

When the supply to the lamp is switched on the heater is energized. Also, the lamp filaments are energized via the starter contact. The heater causes the contacts to part and the choke open-circuits across the tube, so that discharge takes place.

The *glow starter* is the most popular of all the means of starting the discharge. It comprises a pair of open contacts (bimetallic) enclosed in a sealed glass bulb filled with helium gas. This assembly is housed in a metal or plastic canister. *Diagram 120* shows how this type of starter is connected.

When the supply is switched on the helium gas ionizes and heats up, causing the contacts to close, and this energizes the tube filaments. As the contacts have closed, the discharge in the helium ceases, the contacts cool and part, open-circuiting the choke across the tube, and discharge takes place.

In the case of the *quick start* or *instant starter*, starting is achieved by the use of an auto-transformer and an earthed metal strip in close proximity to the tube (*Diagram 121*).

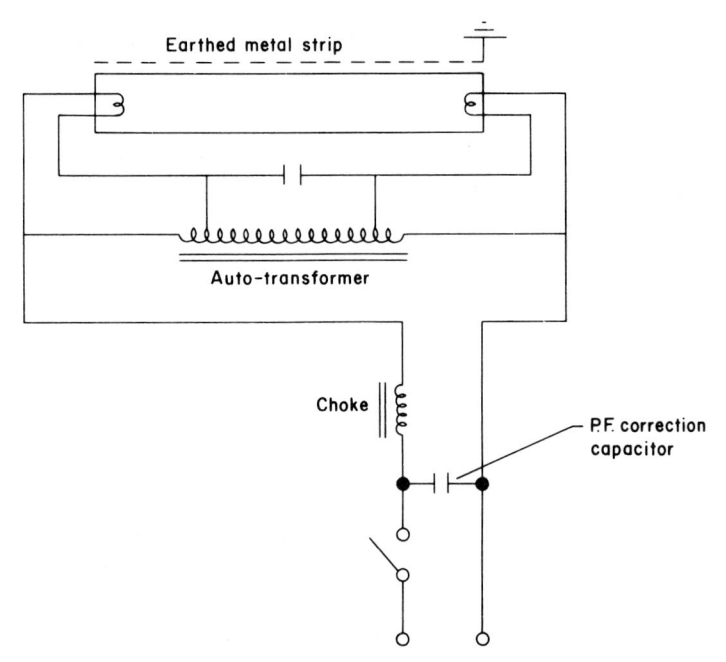

Diagram 121   Quick start or instant starter

When the supply is switched on, mains voltage appears across the ends of the tube, and the small part of the winding at each end of the transformer energizes the filaments, which heat up. The difference in potential between the electrodes and the earthed strip causes ionization, which spreads along the tube.

Fluorescent tube light output
There is a wide range of fluorescent tubes for different applications as *Table 9* indicates.

Table 9

Tube colour	Application
White and warm white	General illumination requiring maximum efficacy, as in drawing offices
Daylight and natural	Any situation requiring artificial light to blend with natural daylight – jewellery, glassware, etc. (main shop areas)
Artificial daylight	Areas where accurate colour matching is carried out
De luxe warm white	Offices and buildings requiring a warm effect, e.g. restaurants, furniture stores
Northlight	Colour-matching areas such as tailors' and furriers'
De luxe natural	Florists', fishmongers', butchers', etc.
Green, gold, blue, red and pink	For special effects

The white tube has the highest efficacy, which for a 2400 mm, 125 W tube is around 70 lm/W.

Points to note (I.E.E. Regulations)
1. If a switch, not designed to break an inductive load, is used to control discharge lighting, it must have a rating not less than twice the steady current it is required to carry, i.e. 10 A switch for a 5 A load.
2. Although a discharge lamp is rated in watts, its associated control gear is highly inductive and therefore the whole unit should have a VA rating. It is on this rating that the current rating of the circuits is calculated. If no technical information is available, a figure of 1.8 is used to calculate the VA rating.

$$\text{i.e. VA rating of 80 W fitting} = 80 \times 1.8$$
$$= 144 \, \text{VA}$$

Lighting and illumination

3. No discharge lighting circuit should use a voltage exceeding 50 kV r.m.s. to earth, measured on open circuit.
4. If a circuit exceeds low voltage and is supplied from a transformer whose rated input exceeds 500 W then the circuit must have protection such that the supply is cut off automatically if short-circuit or earth leakage currents exceed 20% of the normal circuit current.
5. All control equipment including chokes, capacitors, transformers, etc. must either be totally enclosed in an earthed metalwork container or be placed in a ventilated fireproof enclosure. Also, a notice must be placed and maintained on such a container or enclosure, reading 'DANGER – HIGH VOLTAGE'. The minimum size of letters and notice board is as shown in *Diagram 122*.

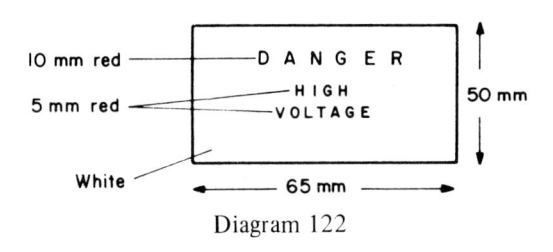

Diagram 122

6. Care must be taken to ensure that the only connection between discharge lamp circuits, operating at a voltage exceeding low voltage, and the mains supply is an earth conductor and/or the earthed neutral conductor of an auto-transformer having a maximum secondary voltage of 1.5 kV.
7. It is important that discharge lighting has a means of isolation from all poles of the supply. This may be achieved in one of the following ways:
    (a) An interlock device, on a self-contained discharge lighting unit, so that no live parts can be reached unless the supply is automatically disconnected (i.e. microswitch on the lid of the luminaire which will disconnect supply to a coil of a contactor when the lid is opened).
    (b) A plug and socket close to the luminaire or circuit which is additional to the normal circuit switch.
    (c) A lockable switch or one with a removable handle or a lockable distribution board. If there is more than one such switch, handles and keys must not be interchangeable.

Lighting and illumination 135

8. Every discharge lighting installation must be controlled by a fireman's switch which will isolate all poles of the supply (it need not isolate the neutral of a three-phase four-wire supply).

9. The fireman's switch should be coloured red and have fixed adjacent to it a notice as shown in *Diagram 123*. The notice should also display the name of the installer and/or maintainer of the installation.

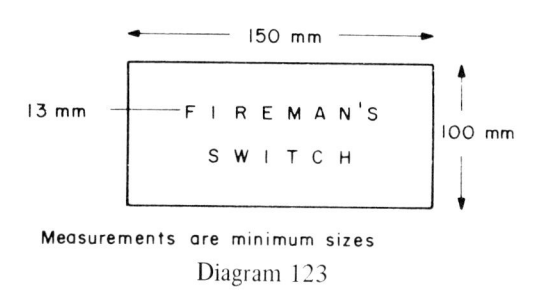

```
         ← 150 mm →
        ┌──────────────┐  ↑
13 mm ──┤  FIREMAN'S   │  │ 100 mm
        │   SWITCH     │  │
        └──────────────┘  ↓
```

Measurements are minimum sizes

Diagram 123

10. The fireman's switch shall have its ON and OFF positions clearly marked, the OFF position being at the top of the switch. The switch should be placed in a conspicuous and accessible position, no more than 2.75 m from ground level.

11. The fireman's switch should be outside and adjacent to the installation for external installations and in the main entrance of a building for interior installations.

12. In general, cables used in discharge lighting circuits exceeding low voltage should be metal-sheathed or armoured unless they are housed in a box sign or a self-contained luminaire or are not likely to suffer mechanical damage.

13. All cables should be supported and placed in accordance with the tables shown in the Regulations.

14. If it is not clear that a cable is part of a circuit operating above low voltage, it should be labelled every 1.5 m as shown in *Diagram 124*.

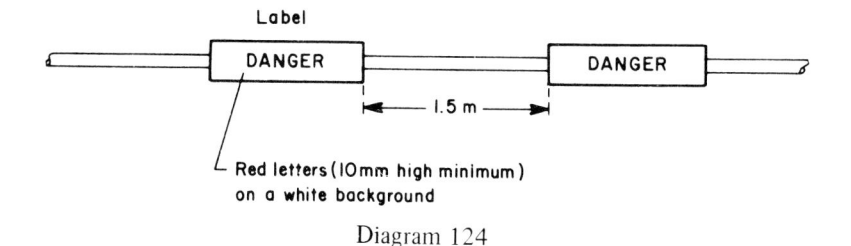

```
           Label
        ┌─────────┐         ┌─────────┐
────────┤ DANGER  ├─────────┤ DANGER  ├────────
        └─────────┘         └─────────┘
                │← 1.5 m →│
              Red letters (10mm high minimum)
              on a white background
```

Diagram 124

136 Lighting and illumination

Questions on Chapter 7

1. Explain what is meant by: (a) 'maintenance factor', (b) 'coefficient of utilization' and (c) 'the inverse square law'.

2. (a) What is meant by 'illuminance'?
   (b) A light source of 850 cd is situated 2.5 m above a work surface. Calculate the illuminance directly below the light and 3 m horizontally away from it (at work surface level).

3. A small workshop 27 m × 17 m requires illuminance at bench level of 130 lx. Two types of lighting are available; (a) 150 W tungsten filament lamps at 13 lm/W, or (b) 80 W fluorescent lamps at 35 lm/W. Assuming that the maintenance factor in each case is 0.8 and that the coefficient of utilization is 0.6, calculate the number of lamps required in each case.

4. With the aid of circuit diagrams explain the difference between a thermal and a glow-type starter for a fluorescent lamp.

5. Show with the aid of a sketch the construction of a high-pressure mercury vapour lamp. Clearly label all parts.

6. Why should a low-pressure sodium vapour lamp be mounted horizontally? What is the most common application for such a lamp? Why is this?

7. Compare the colour and efficacy of tungsten filament, low-pressure sodium and low-pressure mercury vapour lamps.

8. What colour of fluorescent lamp should be used for the following: (a) a butcher's shop, (b) a restaurant and (c) a jeweller's?

9. Explain the purpose of the choke in a discharge lighting circuit.

10. (a) A small supermarket 20 m long by 15 m wide is to be illuminated to a level of 600 lx by 2400 mm 125 W fluorescent lamps having an efficacy of 65 lm/W. The maintenance factor is 0.6 and the coefficient of utilization is 0.85. Calculate the number of fittings required and show their positions on a scale plan.
    (b) Calculate the total current taken by the lighting.

CHAPTER 8

Testing, Inspection and Instruments

Before any completed installation work is connected to the supply system it must be tested and inspected to ensure that it complies with the requirements of the I.E.E. Regulations. It is also important to find out, by testing, the state of a system before any work is carried out.

The following recommended tests shall be carried out where necessary on a completed installation.

1. Continuity of ring final circuit conductors.
2. Continuity of protective conductors.
3. Earth electrode resistance.
4. Insulation resistance.
5. Insulation of site built assemblies.
6. Protection by electrical separation.
7. Protection by barriers or enclosures provided during erection.
8. Insulation of non-conducting floors and walls.
9. Polarity.
10. Earth fault loop impedance.
11. Operation of residual current and fault voltage devices.

These tests, the instruments used, and the allowable values are described in detail in Volume 2.

It is, of course, necessary to test and inspect an installation regularly, ideally every three years. This ensures a good degree of safety, and — in the case of high resistance faults to earth — economy.

Tests on earth-leakage circuit breakers

Where e.l.c.b.s are installed, pushing the test button does *not* prove that the unit will operate correctly under fault conditions. The Regulations recommend that a special test be carried out using a 240 V/45 V step-down transformer, the secondary voltage being applied between neutral and earth. The breaker should trip out instantaneously (*Diagram 125*).

138 *Testing, inspection and instruments*

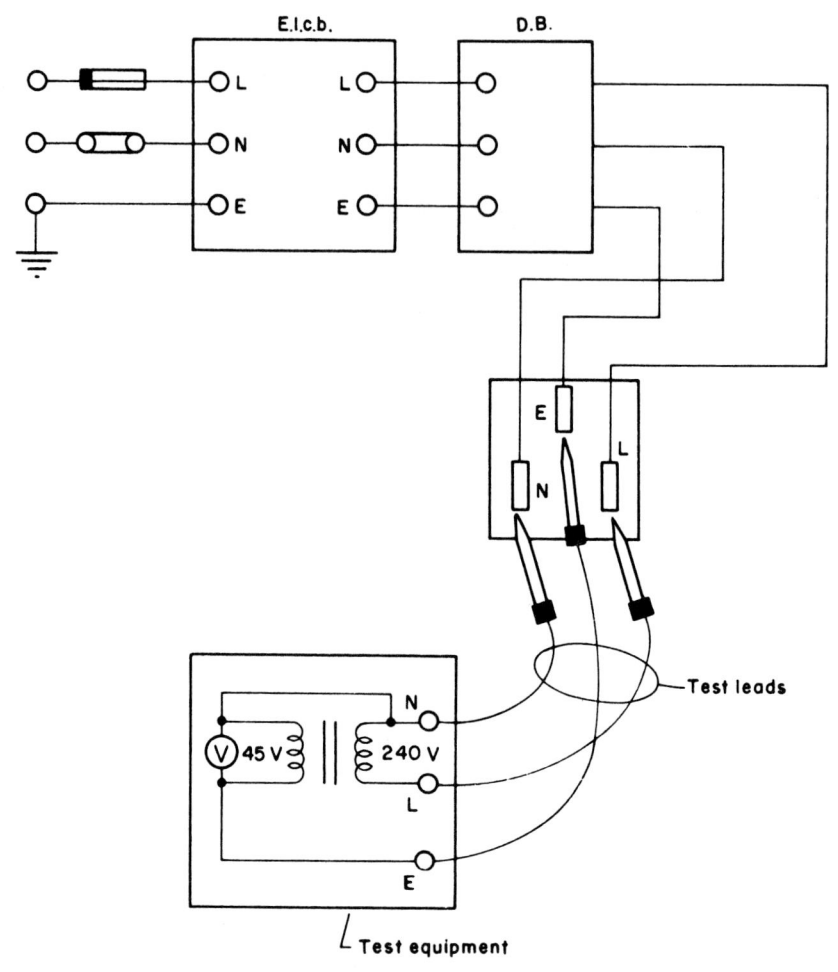

Diagram 125 Testing an earth-leakage circuit breaker. D.B. = distribution board

Measurement of current

In order to measure d.c. currents or voltages, either a moving-coil or a moving-iron instrument may be employed, but for a.c. values a moving-iron instrument must be used. It is often necessary to extend the range of an ammeter to read values of current higher than the instrument's movement is designed for, and for this purpose, shunts or current transformers are used.

Testing, inspection and instruments 139

Ammeter shunts

As *Diagram 126* shows, a shunt is simply a low-value resistor connected in parallel with the instrument.

Example 8.1

A moving-coil ammeter gives full-scale deflection (f.s.d.) at 15 mA. If the instrument resistance is 5 Ω, calculate the value of shunt required to enable the instrument to read currents up to 3 A.

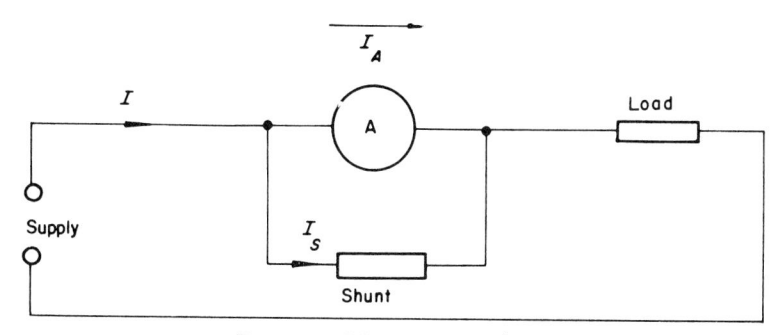

Diagram 126   Ammeter shunt

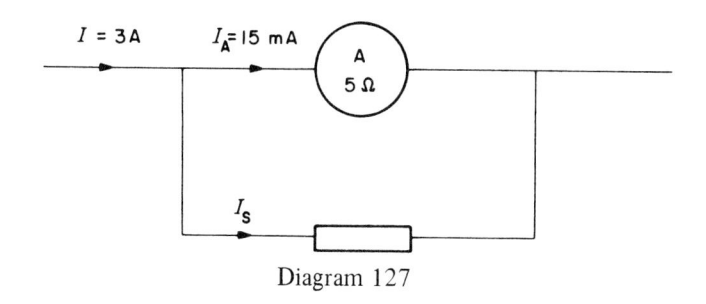

Diagram 127

For full-scale deflection, *Diagram 127* gives:

$$\text{Potential difference across meter} = I_A R_A$$

$$= 15 \times 10^{-3} \times 5$$

$$= 75 \text{ mV}$$

$$\therefore \text{Shunt resistance} = \frac{V_S}{I_S}$$

$$= \frac{75 \times 10^{-3}}{2.985}$$

$$= 0.025 \, \Omega$$

The shunt may be used in conjunction with either a.c. or d.c. instruments. For measuring high a.c. currents, however, a current transformer is used.

$$\therefore \text{Potential difference across shunt} = 75 \, \text{mV}$$

$$\text{Shunt current} = I - I_A$$
$$= 3 - (15 \times 10^{-3})$$
$$= 3 - 0.015$$
$$= 2.985 \, \text{A}$$

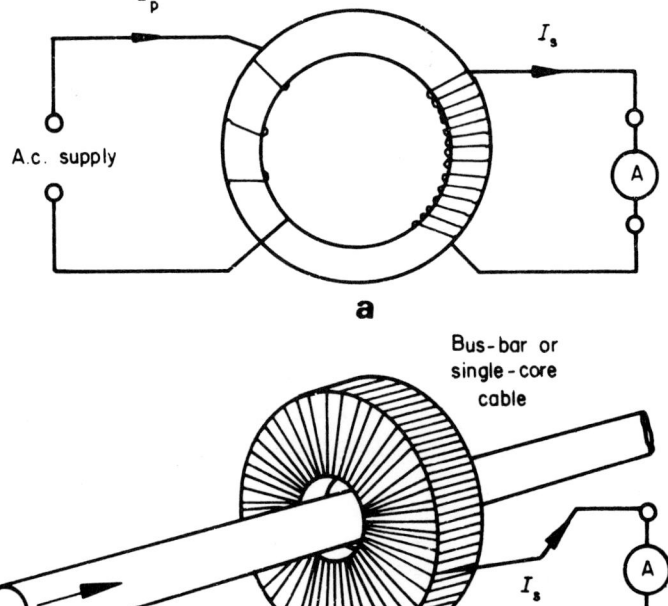

Diagram 128 Current transformers. (a) Wound-type primary; (b) bar-type primary

$$\therefore \text{Shunt resistance} = \frac{V_S}{I_S}$$

$$= \frac{75 \times 10^{-3}}{2.985}$$

$$= 0.025 \, \Omega$$

The shunt may be used in conjunction with either a.c. or d.c. instruments. For measuring high a.c. currents, however, a current transformer is used.

Current transformer

Current transformers (C.T.s) are usually of the wound or bar type shown in *Diagram 128*. As in any transformer, the secondary current will depend on the transformer ratio, i.e.

$$\frac{I_p}{I_s} = \frac{N_s}{N_p}$$

Example 8.2

An ammeter capable of taking 2.5 A is to be used in conjunction with a current transformer to measure a bus-bar current of up to 2000 A. Calculate the number of turns on the transformer.

$$\frac{I_p}{I_s} = \frac{N_s}{N_p}$$

$$\frac{2000}{2.5} = \frac{N_s}{1}$$

$$\therefore N_s = \frac{2000}{2.5}$$

$$= 800 \text{ turns}$$

Great care must be taken when using C.T.s, as high voltages normally associated with high currents will be stepped up on the secondary side, creating a potentially dangerous situation.

Before removing an ammeter or load (burden) from a C.T., the secondary terminals must be shorted out.

Measurement of voltage

As with current measurement, moving-iron and moving-coil instruments are used.

The extension of the range of a voltmeter is achieved by using a multiplier or, for high a.c. voltages, a voltage transformer.

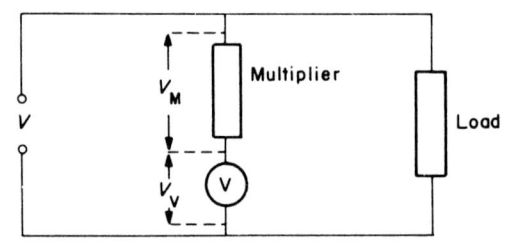

Diagram 129 Voltmeter multiplier

Voltmeter multiplier

A voltmeter multiplier is simply a resistance in series with the instrument as shown in *Diagram 129*.

Example 8.3

A moving-coil instrument of resistance 5 Ω and f.s.d. at 20 mA is to be used to measure voltages up to 100 V. Calculate the value of the series multiplier required.

$$\text{Instrument voltage at f.s.d.} = I_V \times R_V$$

$$= 20 \times 10^{-3} \times 5$$

$$= 0.01 \text{ V}$$

$$\therefore \text{Voltage to be dropped across multiplier} = V - V_V$$

$$= 100 - 0.01$$

$$= 99.99 \text{ V}$$

$$\therefore \text{Value of resistance} = \frac{V_m}{I_m}$$

$$= \frac{99.99}{20 \times 10^{-3}}$$

$$= 4999.5 \text{ Ω}$$

Voltage transformer

A voltage transformer (V.T.) is simply a typical double-wound step-down transformer with a great many turns on the primary and a few on the secondary.

Instruments in general

Multimeters

There are many types of multimeter now available, the more expensive usually giving greater accuracy. They all work on the moving-coil principle and use rectifiers for the d.c. ranges. Shunts, multipliers, V.T.s and C.T.s are switched in or out when ranges and scales are changed by the operator.

Wattmeters

A wattmeter is simply a combination of an ammeter and a voltmeter in one instrument (*Diagram 130*), usually a dynamometer.

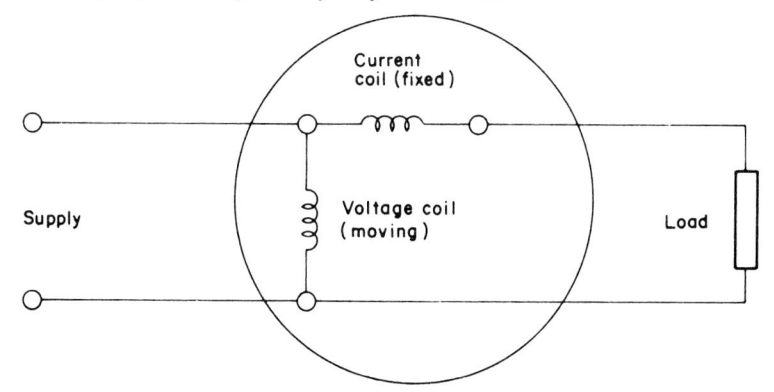

Diagram 130 Wattmeter connection

Diagram 131 shows how high-voltage connections can be made to a wattmeter.

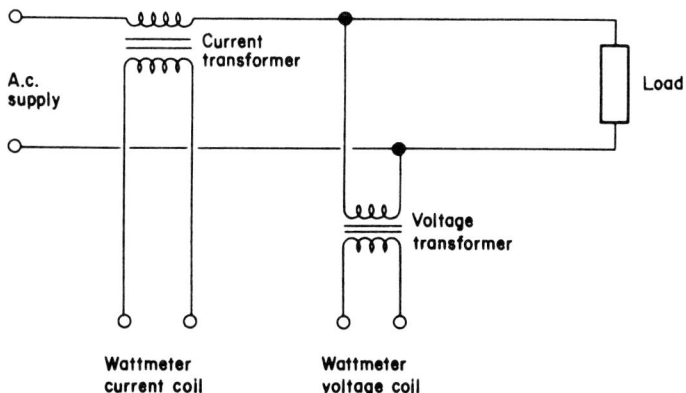

Diagram 131 Use of current transformer and voltage transformer

Tong tester

The tong tester or clip-on ammeter is a variation of the bar-primary current transformer. It consists of an insulated iron core in two parts that can be separated (like tongs), on one end of which is the secondary winding and an ammeter.

The core is clipped round a bus-bar or single-core cable and the current is registered on the ammeter (*Diagram 132*).

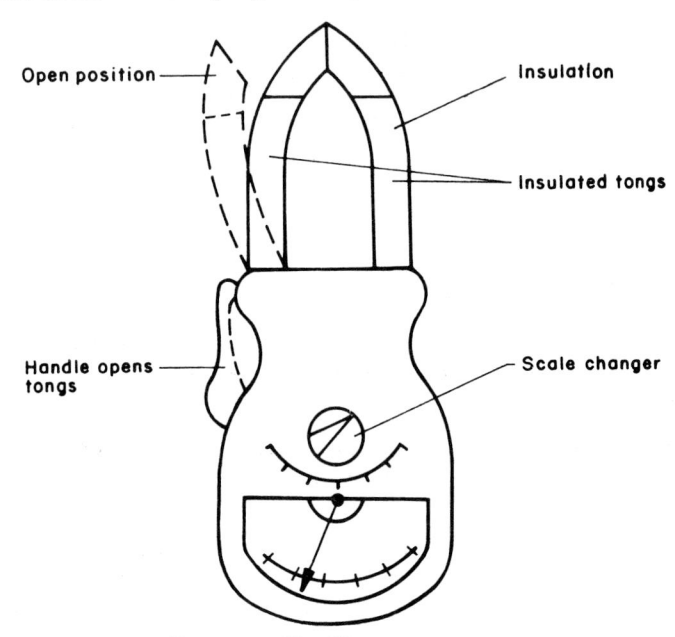

Diagram 132   Clip-on ammeter

Phase rotation indicator

The phase rotation indicator is a simple three-phase induction motor. When connected to a three-phase supply, a disc, connected to the motor, rotates in the direction of the supply sequence. It is used when two three-phase systems are to be paralleled together (*Diagram 133*).

Digital instruments

Recent sudden advances in the field of micro-electronics have enabled many instruments to be developed with a direct digital read-out. They are as yet very expensive but the high degree of accuracy and the minimizing of human reading error make them very attractive.

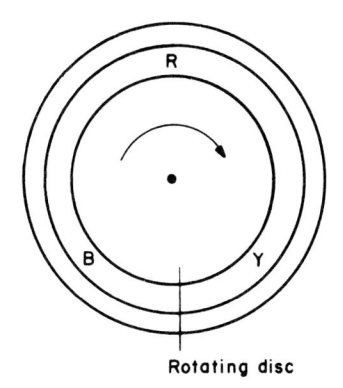

Diagram 133 Phase rotation indicator

Questions on Chapter 8

1. A moving-coil meter, f.s.d. at 15 mA, has a resistance of 5 Ω and is shunted by a resistance of 0.001 Ω. Calculate the maximum current the meter will read.

2. A moving-coil instrument has a resistance of 40 Ω and gives f.s.d. at 4 mA. Calculate the value of the shunts required to give f.s.d. at (a) 0.2 A and (b) 2 A.

3. The movement of an instrument has a resistance of 10 Ω and gives f.s.d. at 5 mA. Calculate the value of shunt required to convert the movement into an ammeter able to read up to 25 A. What value of multiplier will convert the movement to a 0–250 V voltmeter?

4. An ammeter takes a current of 1.5 A for f.s.d. Calculate the number of turns needed on a bar-type primary C.T. if the current to be measured is 450 A.

5. A voltage transformer has a ratio of 2200:1. Calculate the voltage being measured if the voltmeter connected to the secondary of the transformer indicates 5 V.

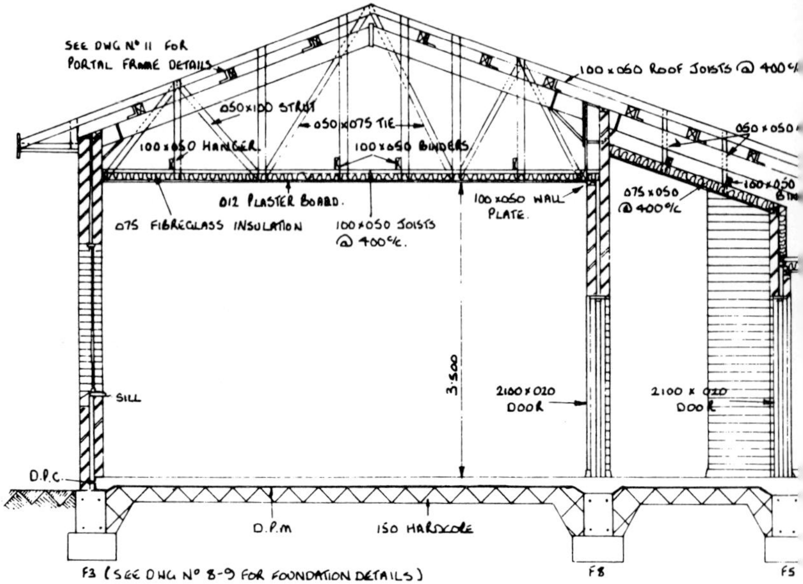

SECTION A-A (SEE DWG No 2) SCALE 1:50

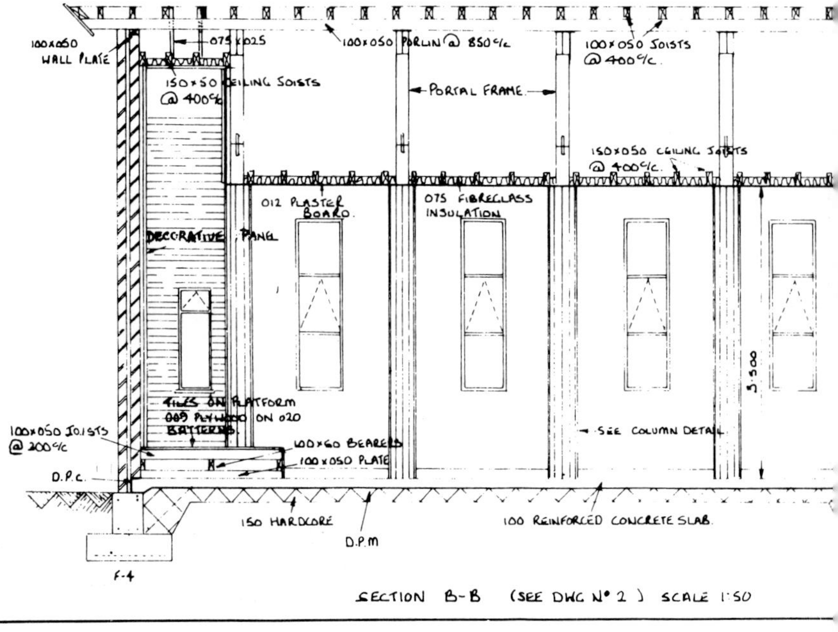

SECTION B-B (SEE DWG N° 2) SCALE 1:50

TYPICAL CHAPEL COLUMN DETAIL (SCALE 1:10)

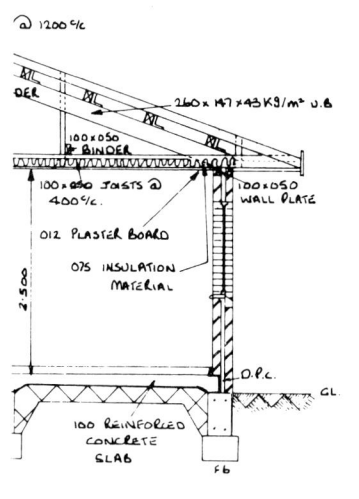

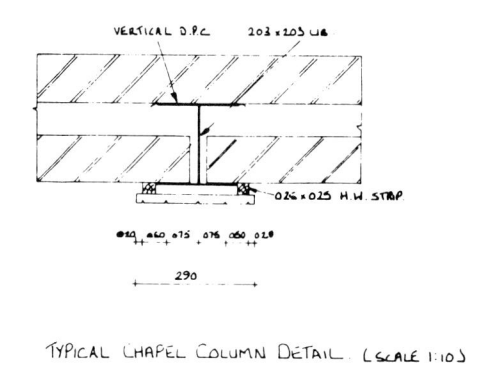

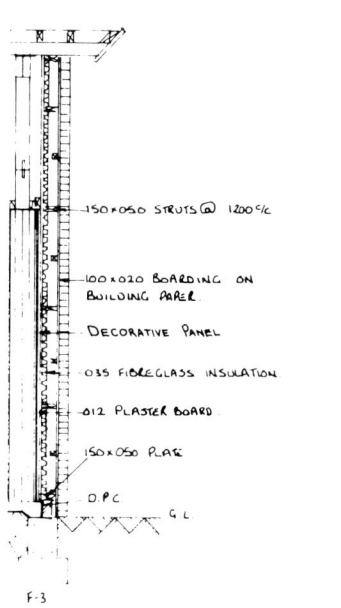

THE CHURCH of JESUS CHRIST of LATTER - DAY SAINTS

SOUTHERN ELEVATION (SCALE 1:100)

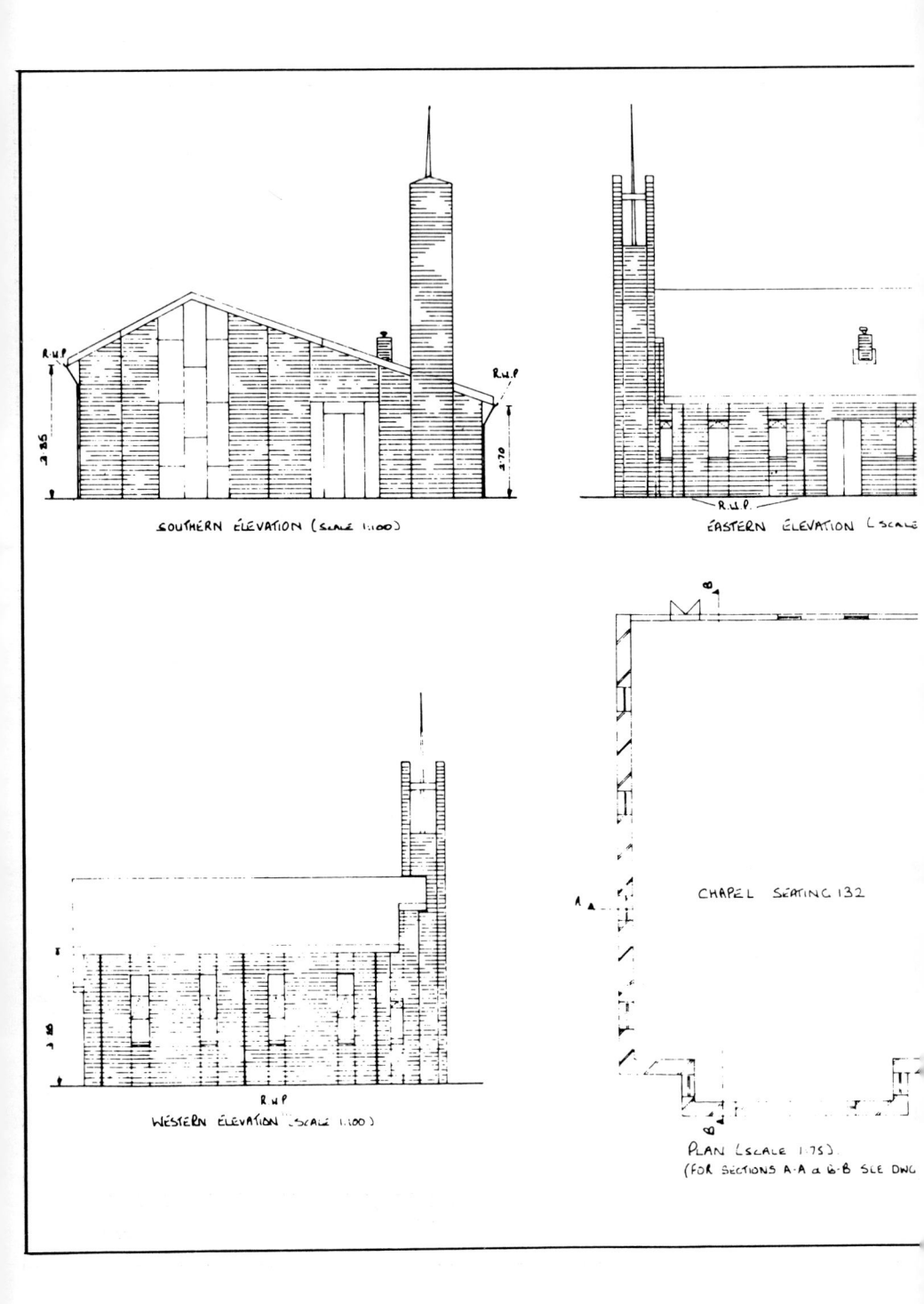

EASTERN ELEVATION (SCALE

WESTERN ELEVATION (SCALE 1:100)

CHAPEL SEATING 132

PLAN (SCALE 1:75)
(FOR SECTIONS A-A & B-B SEE DWG

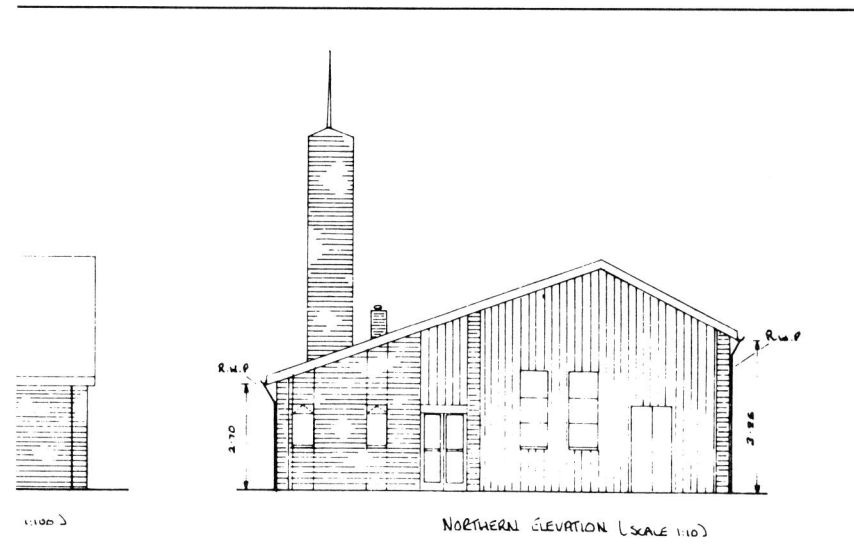

NORTHERN ELEVATION (SCALE 1:10)

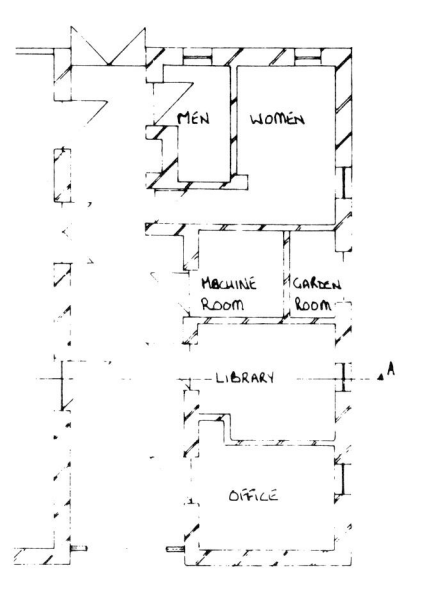

NOTE: DARK BROWN BRICK PANELS ABOVE AND BELOW ALL WINDOWS

REDLAND STONEWOLD TILES TO BE USED ON ROOF

SANDSTONE YELLOW BRICKS TO BE USED THROUGHOUT

MEN
WOMEN
MACHINE ROOM
GARDEN ROOM
LIBRARY
OFFICE

THE CHURCH of JESUS CHRIST of LATTER - DAY SAINTS

Answers to Test Questions

Chapter 2
1. 110 V; 0.95 leading
2. 66.4 μF
3. 5.57 μF
4. 0.787 A at 0.976 lagging
5. 7.8 A at 0.4 lagging
6. 10.8 kVA; 0.82 lagging
7. 0.938 lagging
8. 5.77 A; 110 V
9. 0.8
10. 6 kW; 7.2 kVA

Chapter 3
2. (b) 434 V; 390.6 V; 46.2 A
3. 28 A
4. (b) 8.33 revs/second
6. (b) 2.76%
9. 72 A; 443 μF
10. 118.7 Nm

Chapter 6
1. 231.78 V
2. 234.306 V; 236.256 V
3. 25 mm^2
4. 25 mm^2

Chapter 7
2. (b) 136 lx; 62 lx
3. (a) 64; (b) 45
10. (a) 44; (b) 41.25 A

Chapter 8
1. 75 A
2. (a) 0.816 Ω; (b) 0.0802 Ω
3. (a) 10.002 Ω; (b) 49 990 Ω
4. 300 turns
5. 11 kV

Index

A.c. motors,
  single-phase 56
    capacitor-start 57
    capacitor-start capacitor-run 59
    reactance-start 58
    repulsion-start 60
    resistance-start 58
    shaded-pole 57
    universal or series 60
  three-phase 51
    squirrel-cage induction 53
    synchronous 52
    synchronous-induction 53
    wound-rotor type 54
A.c. theory 6
Agricultural installations 113

Back e.m.f. 39
Balanced three-phase systems 28
Bar charts 4
Bill of quantities 3
British Standards and Codes of Practice 1, 2, 110
Bus-bar trunking 104

Cable selection 114
Concentric cable 98
Contracts 2
Correction factors 115
Cosine rule 122
Current transformers 140

D.c. face-plate starter 46
D.c. generators 49
  separately excited 50

D.c. machines 36
  compound 47
  reversing 49
  series 41
  shunt 44
Daywork 4
Digital instruments 145
Diode 84
Discharge lighting 126
  I.E.E. Regulations concerning 133

Earth-electrode resistance 93
  measurement of 95
Earth-fault loop path 91
Earthed concentric wiring 99
Earthing 91
  I.E.E. Regulations concerning 101
E.l.c.b. (earth-leakage circuit breaker) 97
  testing of 137
Electrical Contractors' Association (E.C.A.) 2
Electricity (Factories Act) Special Regulations 1
Electricity Supply Regulations 1

Fault calculations 95
Fire risk 93, 99
Flammable and explosive situations 110
Fluorescent lamp 130
Frequency of rotor currents 56

Generated e.m.f. 49

151

152 Index

Heat sinks 87

I.E.E. Regulations 1, 80, 90, 101, 126, 133
Impedance triangle 8
Installation systems 103
Installation of motors 63
Instrument shunts and multipliers 138, 142
Inverse-square law 121

Joint Industrial Board (J.I.B.) 2

Light sources 124
Lighting and illumination 120
Loop impedance 92

Measurement of power 32
Mercury vapour lamp 128, 130
Motor enclosures 65ff.
Multimeters 142

'Neon' tube 126
'No-volt' protection 60

Off-peak supplies 108
Oil dashpot 62
Overcurrent protection 61

Parallel earth paths 97
Phase rotation indicator 144
Power factor correction 53
   of a.c. motors 75
Power in a.c. circuits 24, 28
Protective multiple earthing (P.M.E.) 98

Rectification 83

Rectifier output 86
Reverse breakdown voltage 84
Rising mains 104

Semiconductors 83
Shock risk 93, 99
Single-loop motor 38
Slip 54
Smoothing 86
Soil resistivity 95
Specifications 3
Starters
   auto-transformer 63
   direct-on-line 60
   for fluorescent lamps 130
   rotor-resistance 63
   star–delta 61
Synchronous speed 52

Temporary installations 110
Tenders 2, 3
Testing 137
Thermistor 61
Thyristors 87
Tong tester 143
Torque 40
   output and 79
Trade Unions 2, 3
Tungsten filament lamp 125

Variation order 3
Volt drop on radial circuits 106

Wattmeter 143
White meter 107